U.S. Marine Corps Close Combat

Plus

USMC Land Navigation Sergeants Course

Department of Defense

Pentagon
Publishing

DEPARTMENT OF THE NAVY
Headquarters United States Marine Corps
Washington, D.C. 20380-1775

18 February 1999

FOREWORD

1. PURPOSE

Today's Marines operate within a continuum of force where conflict may change from low intensity to high intensity over a matter of hours. Marines are also engaged in many military operations other than war, such as peacekeeping missions or noncombatant evacuation operations, where deadly force may not be authorized. During noncombative engagements, Marines must determine if a situation warrants applying deadly force. Sometimes Marines must decide in a matter of seconds because their lives or the lives of others depend on their actions. To make the right decision, Marines must understand both the lethal and nonlethal close combat techniques needed to handle the situation responsibly without escalating the violence unnecessarily. Marine Corps Reference Publication (MCRP) 3-02B, *Close Combat,* provides the tactics, techniques, and procedures of Marine Corps close combat. It also provides the doctrinal basis for the Marine Corps Close Combat Training Program (MCCCTP).

2. SCOPE

This publication guides individual Marines, unit leaders, and close combat instructors in the proper tactics, techniques, and procedures for close combat. MCRP 3-02B is not intended to replace supervision by appropriate unit leaders and close combat instruction by qualified instructors. Its role is to ensure standardization and execution of tactics, techniques, and procedures throughout the Marine Corps.

3. SUPERSESSION

MCRP 3-02B supersedes Fleet Marine Force Manual (FMFM) 0-7, *Close Combat,* dated 9 July 1993. There are significant differences between the two publications. MCRP 3-02B should be reviewed in its entirety.

4. WARNING

Techniques described in this publication can cause serious injury or death. Practical application in the training of these techniques will be conducted in strict accordance with approved Entry Level Close Combat, Close Combat Instructor (CCI), and Close Combat Instructor Trainer (CCIT) lesson plans. Where serious danger exists, the reac er is alerted by the following:

WARNING

5. CERTIFICATION

Reviewed and approved this date.

BY DIRECTION OF THE COMMANDANT OF THE MARINE CORPS

J. E. RHODES
Lieutenant General, U.S. Marine Corps
Commanding General
Marine Corps Combat Development Command

DISTRIBUTION: 14400006600

OVERVIEW OF CLOSE COMBAT

1. Purpose of Close Combat

Close combat is the physical confrontation between two or more opponents. It involves armed and unarmed and lethal and nonlethal fighting techniques that range from enforced compliance to deadly force. The purpose of close combat is to execute armed and unarmed techniques to produce both lethal and nonlethal results. Unarmed techniques include hand-to-hand combat and defense against hand-held weapons. Armed techniques include techniques applied with a rifle, bayonet, knife, baton, or any weapon of opportunity.

2. Continuum of Force

Marines will find themselves in both combative and noncombative situations. The threat level in these situations can rise and fall several times based on the actions of both Marines and the people involved. The escalation of force stops when one person complies with the demands imposed by another person. This range of actions is known as a continuum of force. Continuum of force is the concept that there is a wide range of possible actions, ranging from voice commands to application of deadly force, that may be used to gain and maintain control of a potentially dangerous situation (MCO 5500.6_, *Arming of Security and Law Enforcement [EE] Personnel and the Use of Force).* Continuum of force consists of five levels that correspond to the behavior of the people involved and the actions Marines use to handle the situation (see the table below). Close combat techniques are executed in levels three, four, and five.

Level One: Compliant(Cooperative)

The subject complies with verbal commands. Close combat techniques do not apply.

Level Two: Resistant (Passive)

The subject resists verbal commands but complies immediately to any contact controls. Close combat techniques do not apply.

Continuum of Force		
Level	**Description**	**Actions**
1	Compliant (Cooperative)	Verbal commands
2	Resistant (Passive)	Contact controls
3	Resistant (Active)	Compliance techniques
4	Assaultive (Bodily Harm)	Defensive tactics
5	Assaultive (Serious Bodily Harm/ Death)	Deadly force
Note: Shading indicates levels in which Marines use close combat techniques.		

Level Three: Resistant (Active)

The subject initially demonstrates physical resistance. Marines use compliance techniques to control the situation. Level three incorporates close combat techniques to physically force a subject to comply. Techniques include—

- Come-along holds.
- Soft-handed stunning blows.
- Pain compliance through joint manipulation and the use of pressure points.

Level Four: Assaultive (Bodily Harm)

The subject may physically attack Marines, but he does not use a weapon. Marines use defensive tactics to neutralize the threat. Defensive tactics include the following close combat techniques:

- Blocks.
- Strikes.
- Kicks.
- Enhanced pain compliance procedures.
- Nightstick blocks and blows.

Level Five: Assaultive (Serious Bodily Harm/Death)

The subject usually has a weapon and will either kill or seriously injure someone if he is not stopped immediately and brought under control. Typically, to control the subject, Marines apply deadly force through the use of a firearm, but they may also use armed and unarmed close combat techniques.

3. Marine Corps Tactical Concepts

Close combat techniques support the following key Marine Corps tactical concepts. The concepts are not standalone ideas but are to be combined to achieve an effect that is greater than their separate sum.

Achieving a Decision

Achieving a decision is important in close combat. An indecisive fight wastes energy and possibly Marines' lives. Whether the intent is to control an opponent through restraint or defend themselves in war, Marines must have a clear purpose before engaging in close combat and act decisively once engaged.

Gaining an Advantage

A basic principle of martial arts is to use the opponent's strength and momentum against him to gain more leverage than one's own muscles alone can generate, thereby gaining an advantage. In close combat, Marines must exploit every advantage over an opponent to ensure a successful outcome. This can include employing various weapons and close combat techniques that will present a dilemma to an opponent. Achieving surprise can also greatly increase leverage. Marines try to achieve surprise through deception, stealth, and ambiguity.

Speed

Marines use speed to gain the initiative and advantage over the enemy. In close combat, the speed and violence of the attack against an opponent provides Marines with a distinct advantage. Marines must know and understand the basics of close combat so they can act instinctively with speed to execute close combat techniques.

Adapting

Close combat can be characterized by friction, uncertainty, disorder, and rapid change. Each situation is a unique combination of shifting factors that cannot be controlled with precision or certainty. For example, a crowd control mission may call for Marines to employ various techniques ranging from nonlethal restraint to more forceful applications. Marines who adapt quickly will have a significant advantage.

Exploiting Success

Typically, an enemy will not normally surrender simply because he was placed at a disadvantage. Marines cannot be satisfied with gaining an advantage in a close combat situation. They must exploit any advantage aggressively and ruthlessly until an opportunity arises to completely dominate the opponent. Marines must exploit success by using every advantage that can be gained.

(reverse blank)

Close Combat

Table of Contents

	Page
Overview of Close Combat	
Chapter 1. Fundamentals of Close Combat	
1. Ranges of Close Combat	1-1
2. Weapons of the Body	1-2
3. Target Areas of the Body	1-3
4. Pressure Points of the Body	1-4
5. Basic Warrior Stance	1-6
6. Angles of Approach and Movement	1-7
7. Balance and Off-Balancing	1-8
8. Falls	1-9
Chapter 2. Lethal and Nonlethal Weapons Techniques	
1. Bayonet Techniques	2-1
2. Nonlethal Rifle and Shotgun Retention Techniques	2-6
3. Nonlethal Handgun Retention Techniques	2-8
4. Firearm Disarmament Techniques	2-12
Chapter 3. Hand-Held Weapons	
1. Fundamentals of Knife Fighting	3-1
2. Knife Fighting Techniques	3-3
3. Weapons of Opportunity	3-6
4. Fundamentals of Combative Stick	3-6
5. Combative Stick Techniques	3-7
6. Blocking Techniques	3-9
7. Unarmed Against Hand-Held Weapons	3-11
8. Counters to Hand-Held Weapon Attacks	3-13
Chapter 4. Strikes	
1. Principles of Punches	4-1
2. Punches	4-2
3. Strikes with the Upper Body	4-4
4. Strikes with the Lower Body	4-10
5. Counters to Strikes	4-12

Page

Chapter 5. Throws

1. Turning Throw 5-1
2. Hip Throw 5-2
3. Leg Sweep 5-3

Chapter 6. Chokes and Holds

1. Types of Chokes 6-1
2. Chokes 6-1
3. Counters to Chokes and Holds 6-5

Chapter 7. Ground Fighting

1. Offensive Ground Fighting 7-1
2. Defensive Ground Fighting 7-3
3. Ground Fighting Chokes 7-5

Chapter 8. Nonlethal Techniques

1. Unarmed Restraints and Manipulation 8-1
2. Nonlethal Baton 8-9

Appendices

A. Pugil Stick Training A-1
B. Safety Precautions During Training B-1

CHAPTER 1

FUNDAMENTALS OF CLOSE COMBAT

This chapter describes all techniques for a right-handed person. However, all techniques can be executed from eittwr side.

The Marine is depicted in camouflage utilities. The opponent is depicted without camouflage.

The fundamentals of close combat include ranges, weapons of the body, target areas of the body, and pressure points of the body. These fundamentals form the basis for all close combat techniques. They provide Marines with a common framework regardless of the type of confrontation or the techniques used. If Marines apply these fundamentals properly in a close combat situation, they may save their lives or the lives of fellow Marines.

1. Ranges of Close Combat

Close combat engagements occur within three ranges: long range, midrange, and close range. During any engagement, these ranges may blur together or may rapidly transition from one to another until either the opponent is defeated or the conflict is resolved.

Long Range

During long range engagements, combatants engage each other with rifles, bayonets, sticks, or entrenching tools. See figure below.

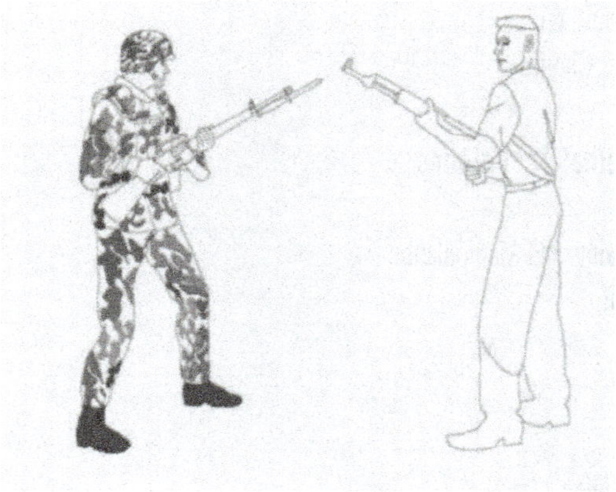

Midrange

During midrange engagements, combatants engage each other with knives, punches, or kicks.

Close Range

During close range engagements, combatants grab each other. Close range engagements also involve elbow strikes, knee strikes, and grappling.

2. Weapons of the Body

Hands and Arms

The hands, forearms, and elbows are the arm's individual weapons. The hands consist of several areas that can be used as weapons: fists, edges of hands, palms, and fingers.

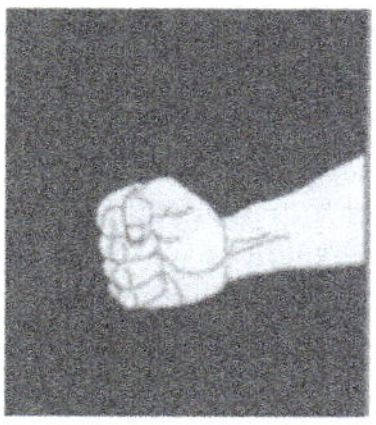

Fists. To minimize injury to the fists, Marines use their fists as weapons to target soft tissue areas such as the throat. The fists' striking surfaces are the first two knuckles of the hands or the meaty portions of the hands below the little fingers.

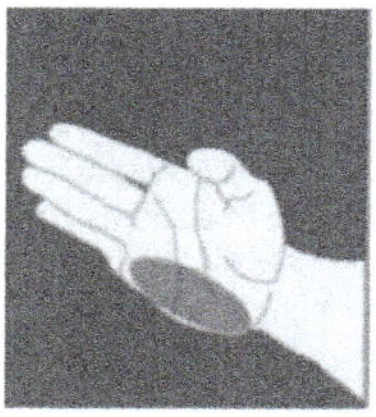

Edge of Hand. Marines use the edge of the hand (knife edge) as a weapon. Marines use the edge of the hand to strike soft tissue areas.

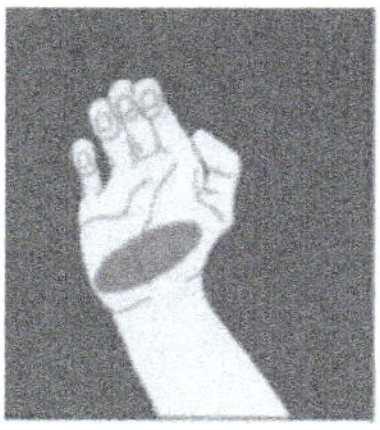

Palms. Because of the palm's padding, Marines use the heels of the palms to strike, parry, and/or block.

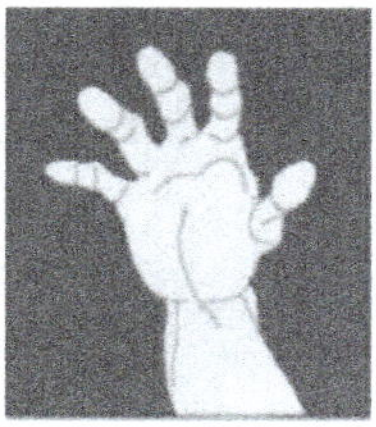

Fingers. Marines use the fingers to gouge, rip, and tear soft tissue areas (e.g., eyes, throat, groin).

Forearms. Marines use the forearms as a defensive tool to deflect or block attacks. Forearms can also be used as striking weapons to damage or break an opponent's joints and limbs. Marines sustain less self-injury when strikes are conducted with the forearms than when strikes are conducted with fists and fingers.

Elbows. Marines use the elbows as striking weapons. Because of the short distance needed to generate power, elbows are excellent weapons for striking during the close range of close combat.

Legs

The legs are more powerful than any other weapon of the body, and they are less prone to injury when striking. The feet are protected by boots and are the preferred choice for striking.

Feet. Marines use the balls of the feet, the insteps, and the toes to kick an opponent. Marines use the cutting edge of the heels and the heels to stomp on an opponent. Marines must be wearing boots when striking with the toes.

Knees. Like elbows, knees are excellent weapons in the close range of close combat. Knee strikes are most effective while fighting close to an opponent where kicks arc impractical. The opponent's groin area is an ideal target for the knee strike if he is standing upright. Knee strikes can deliver a devastating secondary attack to an opponent's face following an initial attack that caused him to bend at the waist.

3. Target Areas of the Body

During close combat, Marines strive to attack the accessible target areas of an opponent's body. The readily accessible areas will vary with each situation and throughout the engagement. The target areas are divided into five major groups: head, neck, torso, groin, and extremities. The figure below illustrates target areas of the body.

Head

The vulnerable regions of the head are the eyes, temple, nose, ears, and jaw. Massive damage to the head kills an opponent.

Eyes. The eyes are excellent targets because they are soft tissue areas that arc not protected by bone or muscle. Attacks to this area may cause the opponent to protect the area with his hands, allowing Marines to execute a secondary attack to other

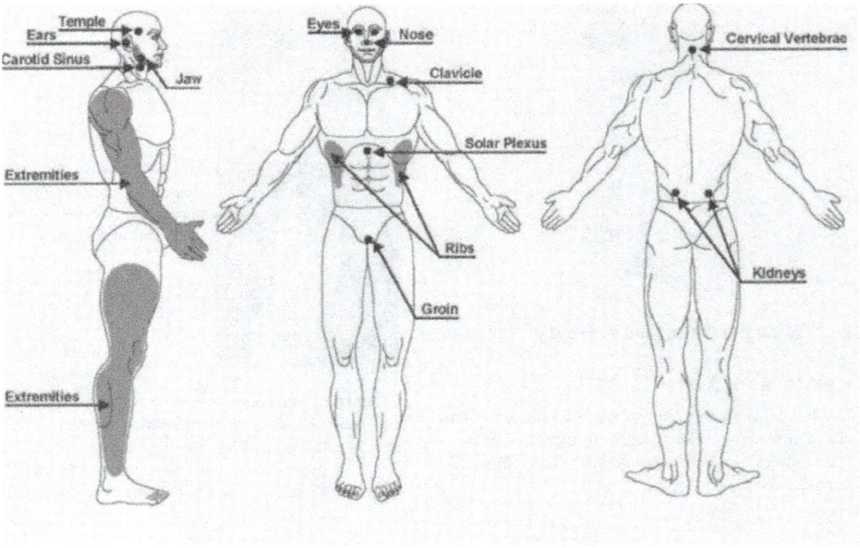

target areas while the opponent uses his hands to protect his eyes.

Temple. The temple is one of the most fragile areas of the head. Powerful strikes to the opponent's temple cause permanent damage and death.

Nose. The nose is very sensitive and easily broken. An attack to this area causes involuntary watering and closing of the opponent's eyes, rendering him vulnerable to secondary attacks. However, through training, individuals can condition themselves to withstand attacks to the nose. Therefore, any attack to the nose must be powerfully delivered.

Ears. Attacks to the ears may cause the eardrum to rupture. But this may not stop or even distract an opponent unless Marines powerfully deliver the strike.

Jaw. The jaw region, when struck forcefully, renders the opponent unconscious. Strikes to the jaw cause painful injuries to the teeth and surrounding tissues (e.g., lips, tongue), but the risk of self-injury is great unless Marines deliver strikes with a hard object such as a helmet, rifle butt, or boot heel.

Neck

The front of the neck, or throat area, is a soft tissue area that is not covered by natural protection. Damage to this region causes the opponent's trachea to swell, closing his airway, which can lead to death.

Carotid Sinus. The carotid sinus is located on both sides of the neck just below the jaw. Strikes to the carotid sinus restrict blood flow to the brain, causing loss of consciousness or death.

Cervical Vertebrae. The cervical vertebrae on the back of the neck, from the base of the skull to the top of the shoulders, contains the spinal cord, which is the nervous system's link to the brain. The weight of the head and the lack of large muscle mass allow damage to the cervical vertebrae and spinal cord. Excessive damage to this area causes pain, paralysis, or death.

Torso

Clavicle. The opponent's clavicle (or collar bone) can be easily fractured, causing immobilization of the arm.

Solar Plexus. Attacks to the opponent's solar plexus or center of the chest can knock the breath out of him and immobilize him.

Ribs. Damage to the opponent's ribs immobilizes him. It may also cause internal trauma.

Kidneys. Powerful attacks to the opponent's kidneys cause immobilization, permanent damage, or death.

Groin

The groin area is another soft tissue area not covered by natural protection. Any damage to this area causes the opponent to involuntarily protect his injured area, usually with his hands or legs. In male opponents, the scrotum is the main target since even a near miss causes severe pain, contraction of the lower abdominal muscles, deterioration of his stance, and possible internal trauma.

Extremities

Rarely will an attack to the opponent's extremities (arms and legs) cause death, but they are important target areas in close combat. Damage to an opponent's joints causes immobilization.

4. Pressure Points of the Body

There are nerves in the human body that, when pressure is applied or when they are struck, allow Marines to control a subject through pain compliance. Marines use pressure points to control an opponent when deadly force is not authorized. They also use pressure points to soften or distract an opponent so a lethal or nonlethal technique can be employed. The figure on page 1-5 illustrates the body's pressure points. Marines execute attacks to pressure points by—

: Rapidly kicking or striking pressure points.
: Slowly applying steady pressure to pressure points.

Infraorbital Nerve

The infraorbital nerve is just below the nose. Marines apply pressure to this nerve with an index finger to control the opponent.

Mastoid Process

The mastoid process is behind the base of the ear and beneath the edge of the jaw. Marines apply inward and upward pressure to this pressure point with the fingers to distract and control the opponent.

Jugular Notch

The jugular notch is at the base of the neck in the notch formed at the center of the clavicle. Marines apply pressure in a quick, stabbing motion with the index finger. Strikes to the jugular notch cause serious damage.

Brachial Plexus (Tie In)

The brachial plexus (tie in) is on the front of the shoulder at the joint. Strikes and pressure applied with the hand are effective on this nerve.

Radial Nerves

Radial nerves are on the inside of the forearms along the radius bones. Strikes and pressure applied with the hand to the radial nerve serve as a softening technique.

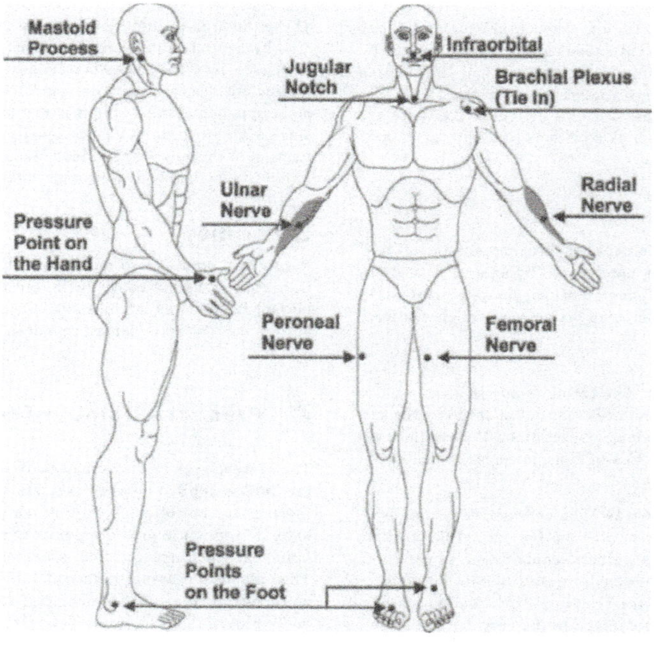

Ulnar Nerve

Ulnar nerves are on the outside of the forearms along the ulnar bones. Strikes and pressure applied with the hand to the ulnar nerve serve as a softening technique.

Pressure Point on the Hand

The hands contain a pressure point on the webbing between the thumbs and index fingers where the two bones of the fingers meet.

To force an opponent to soften or release his grip, Marines apply pressure with their index fingers to this pressure point or strike this pressure point with their fists.

Femoral Nerves

Femoral nerves are on the inside of the thighs along the femur bones. Strikes to the femoral nerve serve as a softening technique.

Peroneal Nerves

Peroneal nerves are on the outside of the thighs along the femur bones. Strikes to the peroneal nerve serve as a softening technique.

Pressure Points on the Feet

There are pressure points on the feet that, when pressure is applied or when they are struck, serve to soften or distract the opponent. Marines apply pressure with the toe, edge, or heel of their boots to the following points:

: The notch below the ball of the ankle.
: The top center of the foot, above the toes.
: The top of the foot where the leg and foot meet.

5. Basic Warrior Stance

Achieving a solid stance is critical to stability and movement throughout any close combat confrontation. The basic warrior stance provides the foundation for stability and movement that is needed to execute close combat techniques. To execute the basic warrior stance, Marines put their feet apart, hands up, elbows in, and chin down.

Feet Apart

Place feet shoulder-width apart.

Keep the head forward and eyes on the opponent, take a half step forward with the left foot, and pivot on the heels so the hips and shoulders are at approximately a 45-degree angle to the right.

Distribute body weight evenly on both legs. Bend the knees slightly.

Hands Up

Curl the fingers naturally into the palm of the hand. Position the thumb across the index and middle fingers. Do not clench the fists. Clenching the fists increases muscular tension in the forearms and decreases speed and reaction time.

Bring the hands up to the face at chin level, with the palms facing each other. Hold the fists up high enough to protect the head, but not so high that they block the field of vision. Ensure continuous eye contact with the opponent.

Elbows In

Tuck the elbows in close to protect the body.

Chin Down

Tuck the chin down to take advantage of the natural protection provided by the shoulders.

6. Angles of Approach and Movement

Marines use movement to control a confrontation and to retain a tactical advantage. Movement increases power and maximizes momentum. By moving around the opponent, Marines gain access to different target areas of the opponent's body. Movement allows Marines to use different weapons of the body and different close combat techniques to attack specific target areas.

Angles of Approach

Marines move anywhere within a 360-degree circle around the opponent to gain a tactical advantage. This circle provides access to different target areas of the opponent's body.

When facing an opponent, Marines move in a 45-degree angle to either side of the opponent. Moving at a 45-degree angle avoids an opponent's strike and puts Marines in the best position to attack the opponent. Marines should avoid being directly in front of an opponent during a confrontation. If a Marine is directly in front of an opponent, the opponent can rely on his forward momentum and linear power to seize the tactical advantage.

Movement

Marines must know how to move in all directions while maintaining the basic warrior stance. During any movement, the legs or feet should not be crossed. Once a movement is completed, the basic warrior stance should be resumed. Maintaining the basic warrior stance protects Marines and puts them in the proper position to launch an attack against an opponent.

Note: Before body movement begins. Marines turn their heads quickly to the new direction. The faster the head turns, the faster the body moves, and the quicker Marines attain visual contact with the opponent.

Forward to the Left. To move forward to the left from the basic warrior stance, Marines-

: Move the left foot forward at a 45-degree angle from the body (approximately 12 to 15 inches), keeping the toe pointed toward the opponent.

Bring the right foot behind the left foot as soon as the left foot is in place. This returns Marines to the basic warrior stance.

Forward to the Right. To move forward to the right from the basic warrior stance, Marines-

Move the right foot forward at a 45-degree angle from the body (approximately 12 to 15 inches).

Bring the left foot, toe pointing toward the opponent, in front of the right foot as soon as the right foot is in place. This returns Marines to the basic warrior stance.

Backward to the Left. To move backward to the left from the basic warrior stance, Marines execute the forward movement in reverse. Marines-

Move the left foot backward at a 45-degree angle from the body (approximately 12 to 15 inches), keeping the toe pointed toward the opponent.

Bring the right foot behind the left foot as soon as the left foot is in place. This returns Marines to the basic warrior stance.

Backward to the Right. To move backward to the right from the basic warrior stance, Marines execute the forward movement in reverse. Marines-

Move the right foot backward at a 45-degree angle from the body (approximately 12 to 15 inches).

Bring the left foot, toe pointing toward the opponent, in front of the right foot as soon as the right foot is in place. This returns Marines to the basic warrior stance.

7. Balance and Off-Balancing

Balance

In any close combat situation, Marines must strive to maintain balance. The last place to be in a close combat situation is on the ground. Marines must maintain a strong base and a low center of balance, their feet must be a shoulder-width apart, and they must stay on their toes to enable quick movement.

Off-balancing

Marines use off-balancing techniques to control an opponent. These techniques are used to throw an opponent to the ground while Marines remain standing, or they are used to put Marines in a position for an offensive attack.

Off-balancing techniques use the opponent's momentum to move or throw him. For example, if the opponent is charging a Marine, the Marine pulls the opponent to drive him to the ground. Likewise, if the opponent is pulling a Marine, the Marine pushes the opponent to drive him to the ground.

Off-balancing techniques also rely on the power generated by the opponent. For example, during combat a Marine may be tired or outnumbered. Depending on the generated energy and momentum of the opponent, the Marine employs off-balancing techniques with very little effort and still provides effective results.

Because off-balancing techniques rely on the momentum and power generated by the opponent, these techniques are particularly effective for Marines who may be outsized by their opponent or lack their opponent's strength.

Angles of Off-balancing

There are eight angles or directions in which an opponent can be off-balanced: forward, rear, right, left, forward right, forward left, rear right, and rear left.

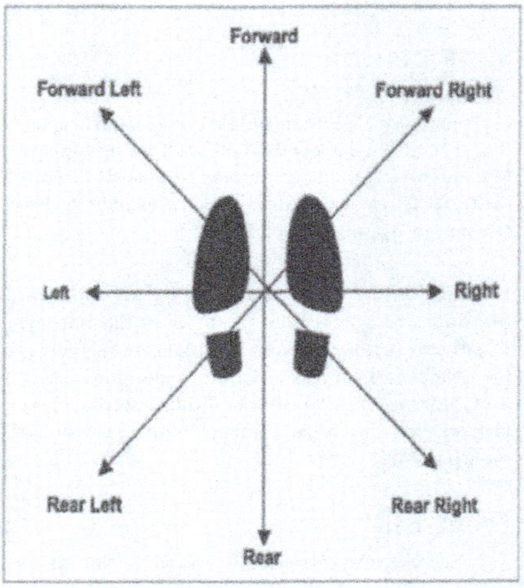

Note: The angles correspond to the Marine's perspective, not the opponent's. Forward, rear, right, and left are straight angles. Forward right, forward left, rear right, and rear left are considered quadrants that are at a 45-degree angle in either direction to the front or rear.

Off-balancing Techniques

Marines off-balance an opponent by pushing, pulling, or bumping the opponent with their hands, arms, or bodies.

To pull, Marines grab an opponent with their hands and drive him forcefully to one of the rear quadrants or to the right or left.

To push, Marines grab the opponent with their hands and drive him forcefully into one of the front quadrants or to the right or left.

Marines execute bumping in the same manner as pushing, but use their shoulders, hips, and legs instead of their hands to off-balance the opponent.

8. Falls

Marines may lose their balance or be thrown to the ground during encounters with an opponent. Marines use falling techniques to absorb the impact of a fall and to quickly return to their feet following an opponent's attack.

Whether falling or being thrown by an opponent, Marines strive to reduce the force of the impact, to prevent serious personal injury, and to increase the chances of survival. Falling techniques use the body's large muscles (back, thighs, buttocks) to protect vital organs and bones from injury and immobilization.

Front Fall

Marines execute a front fall to break a fall on the front. To execute the front fall, Marines-

Bend the elbows and place the palms facing out in a position to spread and absorb the impact of the fall.

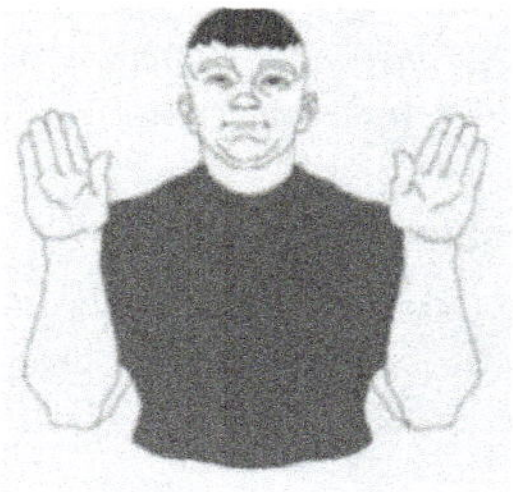

Fall forward, breaking the fall with the forearms and palms. The forearms and hands, down to the fingertips, should strike the ground simultaneous-iy-

Offer resistance with the forearms and hands to keep the head raised off the ground.

Side Fall

Marines execute a side fall to break a fall on the side. To execute the side fall, Marines-

Bring the right arm across the body so the hand is next to the left shoulder with the palm facing inboard.

Fall to the side, breaking the fall with the right arm by slapping the ground and making contact from the shoulder or forearm down to the hand. At the same time, tuck the chin and keep the head raised off the ground. The chin should be tucked to the chest at all times to prevent whiplash.

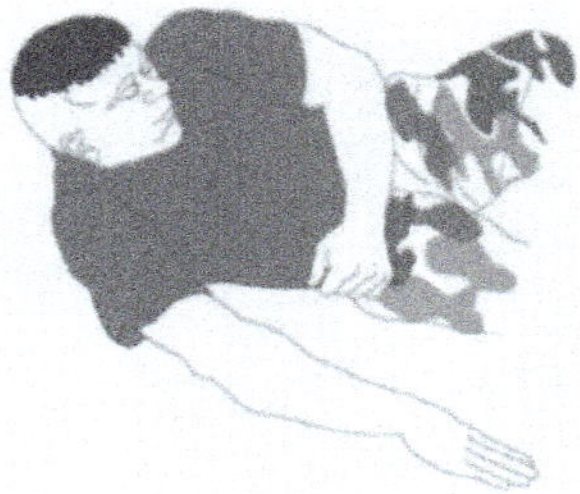

Stretch the right leg out to make contact with the ground and to distribute and absorb the impact. Bend the left leg, allowing the foot to make contact with the ground.

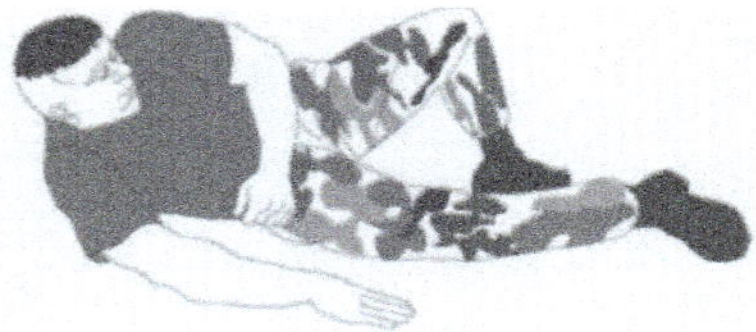

Back Fall

Marines execute a back fall to break the fall when being thrown or falling backward. To execute the backfall, Marines-

Cross the hands in front of the chest and tuck the chin.

Fall backward and slap the ground with the forearms and hands to absorb the impact of the fall and keep the head off the ground.

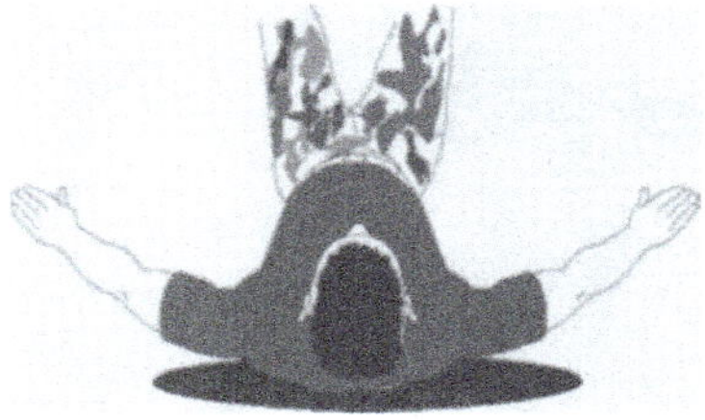

Forward Shoulder Roll

Marines use the forward shoulder roll to break a fall from an opponent's attack and to use the momentum of the fall to get back on their feet quickly. Ideally, Marines execute the forward shoulder roll to a standing position so they can continue fighting. To execute the forward shoulder roll, Marines-

Contact the ground with the back of the right forearm and upper arm. Tuck the chin into the chest.

Roll onto the right shoulder, rolling diagonally across the back to land on the left hip.

Slap the ground with the left arm, absorbing the impact from the shoulder to the hand, palm down.

Keep the left leg straight to absorb as much of the impact as possible. The right leg is bent and the foot hits flat on the ground.

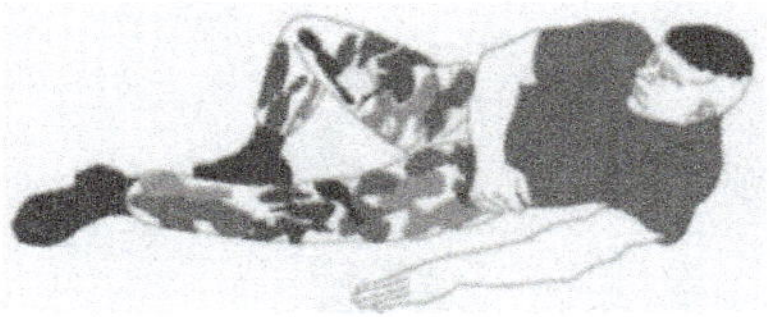

Bend the left leg upon impact to push off with the left knee and leg to a squatting and then a standing position. Forward momentum should carry the Marine to a standing position.

(reverse blank)

CHAPTER 2

LETHAL AND NONLETHAL WEAPONS TECHNIQUES

This chapter describes all techniques for a right-handed person. However, all techniques can be executed from either side.

In drawings, the Marine is depicted in woodland camouflage utilities; the opponent is depicted without camouflage. In photographs, the Marine is depicted in woodland camouflage utilities; the opponent is depicted in desert camouflage utilities.

1. Bayonet Techniques

WARNING

During training, Marines must have bayonets sheathed. Marines use bayonet dummies to practice bayonet techniques. When practicing offensive and defensive bayonet techniques student-on-student, Marines use pugil sticks.

All Marines armed with a rifle carry a bayonet. The bayonet is an effective weapon if Marines are properly trained in offensive and defensive bayonet techniques. An offensive attack, such as a thrust, is a devastating attack that can quickly end a fight. Defensive techniques, such as the block and parry, can deter the opponent's attack and allow Marines to regain the initiative. Through proper training, Marines develop the courage and confidence required to effectively use a bayonet to protect themselves and destroy the enemy. In situations where friendly and enemy troops are closely mingled and rifle fire and grenades are impractical, the bayonet becomes the weapon of choice.

Holding the Rifle

To execute bayonet techniques, Marines hold the rifle in a modified basic warrior stance. All movement begins and ends with the basic warrior stance. To hold the rifle, Marines—

Use an overhanded grasp to grab the small of the rifle's stock. Use an underhanded grasp to grab the handguards of the rifle.

Lock the buttstock of the rifle against the hip with the right forearm.

Orient the blade end of the rifle toward the opponent.

Offensive Bayonet Techniques

Straight Thrust. Marines use the straight thrust to disable or kill an opponent. It is the most deadly offensive technique because it causes the most trauma to an opponent. Target areas are the opponent' s throat, groin, or face. The opponent's chest and stomach are also excellent target areas if not protected by body armor or combat equipment. To execute the straight thrust, Marines—

Lift the left leg and lunge forward off the ball of the right foot while thrusting the blade end of the weapon forward, directly toward the opponent.

Retract the weapon and return to the basic warrior stance.

Slash. Marines use the slash to kill an opponent or to create an opening in his defense. The target area is the opponent's neck. To execute the slash, Marines—

Extend the left hand back toward the left shoulder.

Thrust the left hand forward and swing it to the right, bring the right hand back toward the hip, and turn the cutting edge of the blade toward the opponent's neck. The movement is a slashing motion so the blade cuts across the opponent's neck.

Horizontal Buttstroke. Marines use the horizontal buttstroke to weaken an opponent's defenses, to cause serious injury, or to set him up for a killing blow. Target areas are the opponent's head, neck, and legs. To execute the horizontal buttstroke, Marines—

Step forward with the right foot and drive the right hand forward. Rotate the hips and shoulders into the strike. Move the left hand back toward the left shoulder.

Strike the opponent with the butt of the weapon.

Vertical Buttstroke. Marines use the vertical buttstroke to weaken an opponent's defenses, to cause serious injury, or to set him up for a killing blow. Target areas are the opponent's groin and face. To execute the vertical buttstroke, Marines—

Step forward with the right foot and drive the right hand straight up.

Pull the left hand back over the left shoulder.

Strike the opponent with the butt of the weapon.

Smash. Marines use the smash as a follow-up technique to the horizontal or vertical buttstroke, particularly if they missed the target. The target area is the opponent's head. To execute the smash following a buttstroke, Marines-

Step forward with the right foot and place the blade end of the weapon over the left shoulder and elevate the right elbow above the shoulders.

Drive the arms straight forward, striking the opponent with the butt of the weapon.

Defensive Bayonet Techniques

Parry. Marines use a parry as a defensive technique to redirect or deflect an attack. A parry is a slight redirection of a linear attack by an opponent; e.g., a straight thrust or a smash. To execute the parry, Marines-

Use the bayonet end of the rifle to redirect the barrel or bayonet of the opponent's weapon.

Lock the weapon against the hip with the right forearm.

Rotate to the right or left, moving the bayonet end of the rifle to parry the opponent's attack. Rotation is generated from the hips.

Redirect or guide the opponent's weapon away from the body by exerting pressure against the opponent' s weapon.

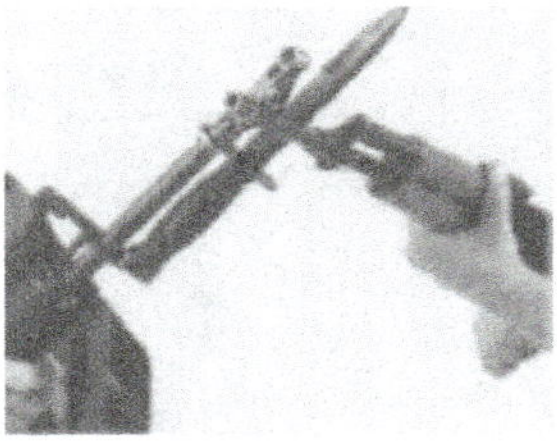

High Block. Marines execute a high block against a vertical attack coming from high to low. To execute the high block, Marines—

Thrust the arms up forcefully at approximately a 45-degree angle from the body. The weapon should be over the top of the head and parallel to the ground. The elbows are bent, but there is enough muscular tension in the arms to absorb the impact and deter the attack.

Low Block. Marines execute the low block

against a vertical attack coming from low to high. To execute the low block, Marines—

Thrust the arms down forcefully at approximately a 45-degree angle from the body. The weapon

should be at or below the waist and parallel to the ground. The elbows are bent, but there is enough muscular tension in the arms to absorb the impact and deter the attack.

Left and Right Block. Marines execute a left or right block against a horizontal buttstroke or a slash. To execute the left or right block, Marines-

Thrust the arms forcefully to the right or left, holding the rifle vertically in the direction of the attack. The elbows are bent, but there is enough muscular tension in the arms to absorb the impact and deter the attack.

Counter Action Following the Block. After deflecting an opponent's attack with a block, Marines counter with a slash or a horizontal buttstroke to regain the initiative. However, the objective in any bayonet fight is to thrust forward with the blade end of the weapon to immediately end the fight.

Group Strategy

On occasion, Marines may engage an opponent as a member of a group or numerous opponents by one's self or as a member of a group. By combining bayonet fighting movements and simple strategies, Marines can effectively overcome their opponent or opponents.

Offensive Strategy: Two Against One. If two bayonet fighters engage one opponent, the fighters advance together.

Fighter 1 engages the opponent while fighter 2 swiftly and aggressively attacks the opponent's exposed flank and destroys the opponent.

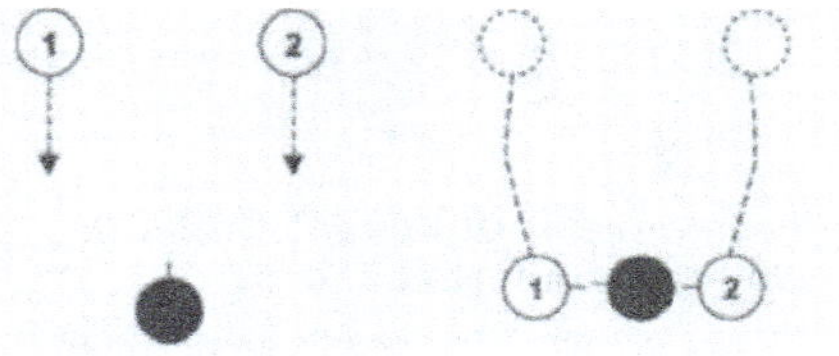

Offensive Strategy: Three Against Two. If three bayonet fighters engage two opponents, the fighters advance together keeping their opponents to the inside.

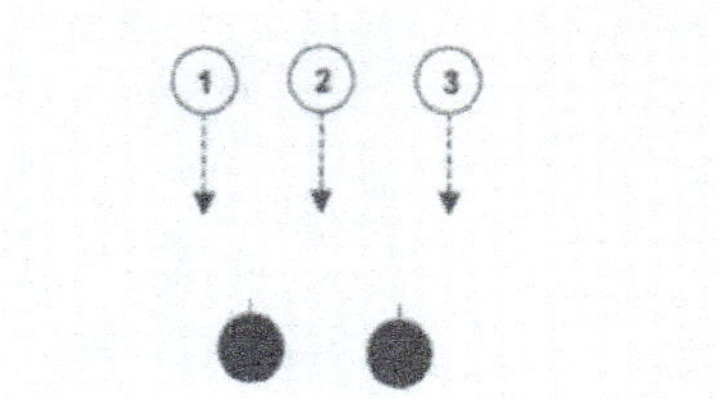

Fighters 1 and 3 engage opponents. Fighter 2 attacks the opponent's exposed flank engaged by fighter 1 and destroys the opponent.

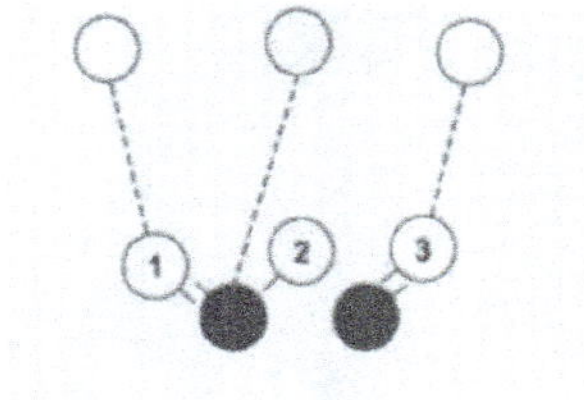

Fighters 1 and 2 turn and attack the exposed flank of the opponent engaged by fighter 3 and destroy the opponent.

Defensive Strategy: One Against Two. If a fighter is attacked by two opponents, the fighter immediately positions himself at the flank of the nearest opponent and keeps that opponent between himself and the other opponent.

Using the first opponent's body as a shield against the second opponent, the fighter destroys the first opponent quickly before the second opponent moves to assist.

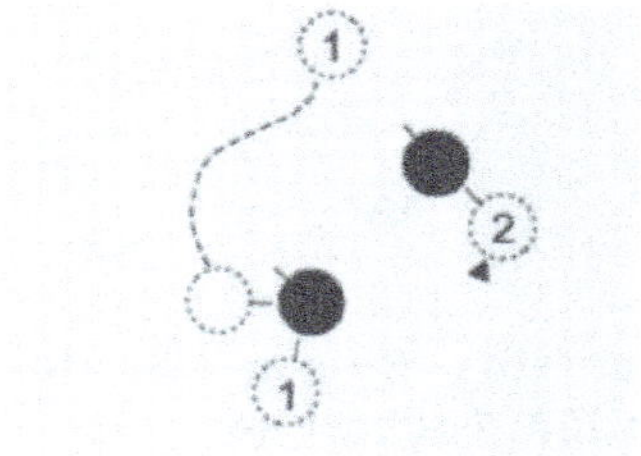

Then, the fighter engages and destroys the second opponent.

Defensive Strategy: Two Against Three. If two fighters are attacked by three opponents, the fighters immediately move to the opponent's flanks.

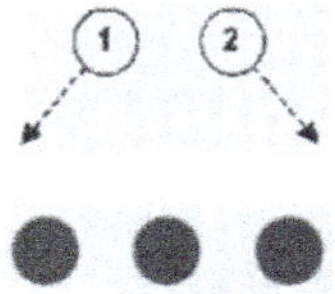

Fighters 1 and 2 quickly attack and destroy their opponents before the third opponent closes in.

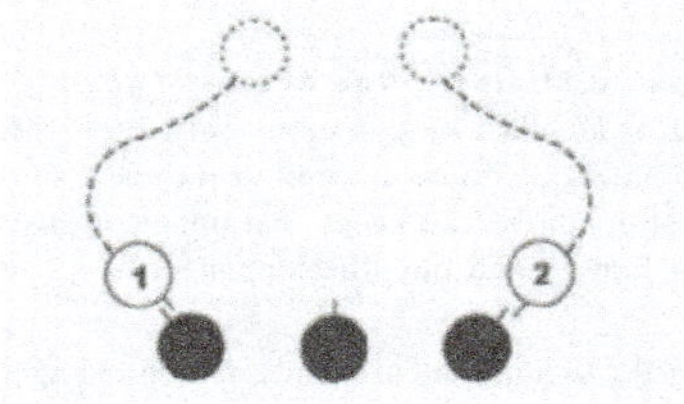

Fighter 1 engages the third opponent while fighter 2 attacks the opponent's exposed flank and destroys the opponent.

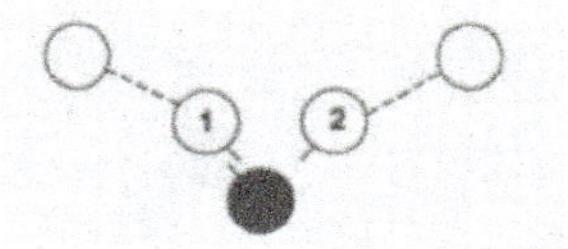

2. Nonlethal Rifle and Shotgun Retention Techniques

Most Marines are armed with the M16A2 service rifle. Marines are taught to keep their weapon with them at all times. Marines must be constantly alert to their surroundings and the people moving in and around their environment. Marines may be confronted by an individual who tries to take theii weapons. If this happens, Marines should not struggle with the individual. To retain positive control of their weapons, Marines must understand and apply weapons retention techniques, otherwise known as armed manipulation. The fol lowing techniques can be used with either the rifle or the shotgun.

Blocking Technique

To execute a blocking technique, Marines-

Stand in a defensive posture.

Use the weapon to block the opponent by thrusting it out firmly, with the elbows bent. Do not try to hit the opponent with the rifle. The rifle is used as a barrier.

Technique if Opponent Grabs Weapon Underhanded

Usually an opponent will try to grab the weapon or block it as an instinctive action. If the opponent uses an underhand grab to seize the handguards of the weapon, Marines—

Trap the opponent's closest finger(s) above the knuckle with the thumb so he cannot release his grip-

Apply bone pressure on the opponent's finger to initiate pain compliance.

Rotate the barrel of the weapon up or down quickly while maintaining pressure on the opponent's hand. At the same time, quickly pivot to off-balance the opponent.

Technique if Opponent Grabs Weapon Overhanded

If the opponent uses an overhand grab to seize the weapon overhanded, Marines—

Trap the opponent's finger to hold his hand in place.

Rotate the barrel to place it across the opponent's forearm and apply downward pressure. This action is similar to an armbar.

Technique if the Opponent Grabs the Muzzle

If the opponent grabs the muzzle of the weapon, Marines—

Rotate the muzzle quickly in a circle motion, Slash downward with the muzzle to release the opponent's grip.

Butt Strikes

Strikes with the butt of the weapon control or ward off an attacker. During any of the retention techniques, Marines use the heel or cutting edge of the weapon to deliver butt strikes to the inside or the outside of the opponent's thigh. The inside butt strike targets the femoral nerve as illustrated in the photo to the left.

The outside butt strike targets the peroneal nerve as illustrated in the photo to the left. Strikes can be made to the outside or inside of the thighs. If a strike to one side of the thigh misses, Marines follow back through with the butt of the weapon on the other side of the thigh.

Off-balancing Techniques

Marines apply off-balancing techniques to throw an opponent to the ground and retain possession of the weapon.

If the opponent grabs the weapon and pushes, Marines should not push on the weapon. They should—

Move with the momentum and movement of the

opponent by pivoting in the direction of the movement by stepping back.

Throw the opponent to the ground with a quick jerking movement by lowering the mu/./le and swinging the butt of the weapon.

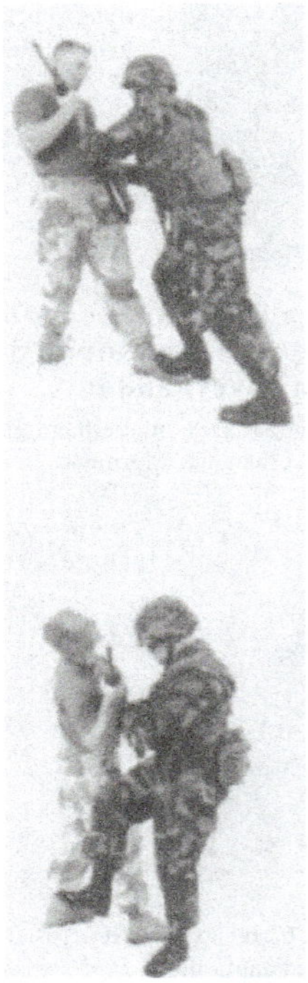

If the opponent grabs the weapon and pulls, Marines—

Step on the opponent's foot and push forward to off-balance him and drive him to the ground.

Sweep the opponent's feet out from under him by hooking his leg with the leg and kicking backward.

3. Nonlethal Handgun Retention Techniques

Many Marines are armed with the M9 service pistol. Marines must keep their weapons in their possession at all times. Marines must be constantly alert to their surroundings and the people moving in and around the environment. Marines may be confronted by an individual who tries to take their weapons. To retain positive control of the weapon, Marines must understand and apply handgun retention techniques.

Blocking Technique

Marines perform the following blocking techniques if an opponent attempts to grab their pistol in the holster. Marines—

Place the body between the weapon and the opponent by immediately pivoting so the weapon is away from the opponent.

Step back and away from the opponent while placing the hand on the pistol grip.

Extend the left hand and block, deflect, or strike the opponent's arm.

Armbar Technique

Marines use the armbar technique when an opponent uses his right hand to grab their pistol in the holster. To execute the armbar technique, Marines—

Trap the opponent's right hand by grasping the opponent's wrist or hand with the right hand and applying pressure against the body.

Step back with the right foot and pivot sharply to the right to be next to the opponent. Always pivot in a direction that keeps the weapon away from the opponent.

Straighten the opponent's arm to apply an armbar. The arm should be straight across the torso.

Continue pivoting to the right while pulling back on the opponent's shoulder. This action may break the opponent's arm.

Place the left hand across the opponent's face and apply pressure back and down to take the oppo-

nent to the ground. Pressure applied beneath the nose or on the trachea is equally effective.

Wristlock Technique

Marines use the wristlock technique when an opponent grabs their pistol while it is in the holster with his right hand. To execute the w ristlock technique, Marines—

Grasp the opponent's wrist or hand with the right hand and apply pressure against the body.

Step back and away from the opponent and pivot to the right so the weapon is away from the opponent. Always pivot in a direction that keeps the weapon away from the opponent.

Reach over the opponent's arm with the left hand and grab his hand, applying pressure against his arm with the left forearm. Execute a wristlock.

Incorporate the second hand into the wristlock and, stepping back with the left foot, pivot to the left.

Execute a two-handed wristlock by exerting downward pressure with the thumb and rotating his hand to the left.

Continue to pivot to off-balance the opponent and drive him to the ground.

Note: If the opponent grabs the pistol with his left hand, Marines execute the wristlock with one hand and step in toward the opponent, rather than

away from the opponent.

Softening Techniques

Handgun retention techniques use softening techniques applied to pressure points. Bone pressure

and strikes with the hands (i.e., hammer fist), knees, and feet are also effective softening techniques.

Pressure Points. Marines use pressure point techniques to get the opponent to loosen his grip. Marines use their finger tips to apply pressure to the webbing between the index finger and thumb, the jugular notch, and the brachial plexus tie in. The following figure illustrates pressure applied to the brachial plexus tie in.

Bone Pressure. Bone pressure is the application of pressure on a bone against a hard object to initiate pain compliance. To apply bone pressure, Marines use their hand to trap the opponent's hand on the weapon. Marines apply a slow, steady pressure to the opponent's hand and fingers until his grip is softened or he releases his hold.

Strikes. If it is difficult to apply a retention technique, Marines employ strikes or kicks to force the opponent to loosen his grip. Strikes to the

eyes, the arms (radial nerve), or shoulder (brachial plexus tic in) soften the opponent's grip on the weapon.

Kicks and knee strikes to the peroneal nerve, the femoral nerve, or the groin arc effective because the opponent is typically unprepared to counter the strike.

Stomping on the top of the opponent's foot may distract him or loosen his grip on the weapon.

4. Firearm Disarmament Techniques

Marines use firearm disarmament techniques during a close-range confrontation if they are un-

armed and the opponent has a firearm (pistol). These techniques are equally effective if Marines are armed but do not have time to withdraw and present the weapon. The goal of firearm disarmament techniques is to gain control of the situation

so Marines gain the tactical advantage. The goal is not necessarily to get control of the opponent's weapon.

Pistol to the Front

This technique is used when Marines are unarmed and the opponent has a pistol pointing at their front (e.g., head, chest). The technique is the same if the opponent sticks the pistol under the Marine's chin. To execute the counter when an opponent is pointing a pistol toward the front of a Marine, Marines—

Place the hands close to the weapon, about chest high, palms out.

Use the left hand to grab the opponent's forearm and push the opponent's hand with the pistol away to clear the body from in front of the weapon. At the same time, rotate the right shoulder back to clear the body from the weapon.

Maintain control of the opponent's arm.

Grasp the weapon with the right hand by placing the thumb underneath the pistol and the fingers over top of the pistol.

Keep the right hand wrapped tightly around the muzzle and quickly rotate the pistol in the opponent's hand so the muzzle is facing the opponent.

Grasp and pull the opponent's wrist or forearm away from the body while rotating the weapon.

Rotate the weapon toward the opponent while pulling it up and back and out of the opponent's grasp.

Pistol to the Rear

This technique is performed when Marines are unarmed and the opponent has a pistol pointing to the back of the Marine's head. To execute the counter when the opponent is pointing a pistol to the rear, Marines—

Place the hands close together about chest high, palms out.

Step back with the left foot, pivoting on the right foot so the side is against the opponent's front. This action clears the body from the weapon's line of fire. Keep the left hand up.

Pivot on the left foot to face the opponent and, at the same time, raise the left elbow and reach over the top of the opponent's arm with the left arm.

Wrap the left arm tightly around the opponent's arm above his elbow to control it.

Push on the opponent's shoulder with the right hand while pulling up with the left arm to achieve an armbar. This action releases the opponent's grip on the weapon. If necessary, the opponent can be taken to the ground with a leg sweep.

Note: To execute this technique, the weapon must be close to or touching the rear of the Marine. If the weapon is too far away from the body, this technique would be difficult to execute or it would be ineffective.

CHAPTER 3

HAND-HELD WEAPONS

This chapter describes all techniques for a right-handed person. However, all techniques can be executed from either side.

In drawings, the Marine is depicted in woodland camouflage utilities; the opponent is depicted without camouflage. In photographs, the Marine is depicted in woodland camouflage utilities; the opponent is depicted in desert camouflage utilities.

Marines must know how to defend against attacks when an opponent is either unarmed or armed with a held-held weapon. This chapter addresses the combative use of knives, specific weapons of opportunity, and sticks. However, virtually anything can be used as a hand-held weapon.

1. Fundamentals of Knife Fighting

Marines must be trained in knife fighting techniques. Marines experienced in offensive knife techniques can cause enough damage and massive trauma to stop an opponent. When engaged against each other, experienced knife fighters employ various maneuvers and techniques that are specific to knife fighting. Seldom, if ever, will Marines engage an opponent in a classical knife fight.

Note: When armed with a rifle, Marines are issued a bayonet. When armed with a pistol, Marine are issued a combat knife.

Angles of Attack

There are six angles from which an attack with a knife can be launched:

- Vertical strike coming straight down on an opponent.
- Forward diagonal strike coming in at a 45-degree angle to the opponent.
- Reverse diagonal strike coming in at a 45-degree angle to the opponent.
- Forward horizontal strike coming in parallel to the ground.
- Reverse horizontal strike coming in parallel to the ground.
- Forward thrust coming in a straight line to the opponent.

Target Areas of the Body

During any confrontation, the parts of the opponent's body that are exposed or readily accessible will vary. The goal in a knife fight is to attack the body's soft, vital target areas that are readily accessible (e.g, the face, the sides and front of the neck, the lower abdomen [or groin]).

Neck. Carotid arteries, located on either side of the neck, are good target areas because they are not covered by body armor or natural protection.

Lower Abdomen (or Groin). The lower abdomen (or groin region) is a good target area because it is not covered by body armor.

Heart. The heart, if not covered by body armor, is an excellent target which, if struck, can prove fatal in a matter of seconds or minutes.

Secondary Targets. There are secondary target areas that will cause substantial bleeding if an artery is severed. These target areas are not immediately fatal, but can become fatal if left unattended. Attacks to—

The legs can cause a great deal of trauma and prove fatal. For example, the femoral artery located on the inside of the thigh is a large artery which, if cut, will cause extensive blood loss.
The brachial artery, located between the biceps and triceps on the inside of the arm, can cause extensive bleeding and damage.
The arm's radial and ulnar nerves can cause extensive bleeding and damage.

Movement

Marines can move anywhere within a 360-degree circle around the opponent. This allows accessibility to different target areas of the opponent's body. Marines should avoid being directly in front of an opponent because the opponent can rely on his forward momentum to seize the tactical advantage. If Marines face an opponent, movement is made in a 45-degree angle to either side of the opponent. This angle avoids an opponent's strike and places Marines in the best position to attack an opponent.

Wearing the Combat Knife

Marines must wear the combat knife where it is easily accessible and where it can best be retained. It is recommended the combat knife be worn on the weak side hip, blade down, sharp edge facing forward. Marines can place it behind the magazine pouch where it is easily accessible to them, but not easily grabbed by an opponent.

Grip

The grip on the knife should be natural. Marines grasp the knife's grip with the fingers wrapped around the grip naturally as it is pulled out of its sheath. This is commonly known as a hammer grip. The blade end of the knife is always facing the opponent.

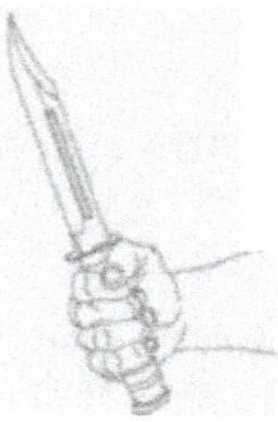

Stance

Marines use the basic warrior stance as the foundation for knife techniques. The left hand is a vertical shield that protects either the ribs or the head and neck. The right elbow is bent with the blade pointing forward toward the opponent's head. This position serves as an index point, where all techniques are initiated.

Principles of Knife Fighting

The following are key principles of knife fighting:

Execute movements with the knife blade within a box, shoulder-width across from the neck down to the waistline. The opponent has a greater chance of blocking an attack if the blade is brought in a wide, sweeping movement to the opponent.

Close with the opponent, coming straight to the target.

Move with the knife in straight lines.

Point the knife's blade tip forward and toward the opponent.

Apply full body weight and power in each of the knife techniques. Full body weight should be put into the attack in the direction of the blade's movement (slash or thrust).

Apply constant forward pressure with the body and blade to keep the opponent off-balanced.

2. Knife Fighting Techniques

Slashing Techniques

Marines use slashing techniques to close with an enemy. Slashing techniques distract the opponent or damage the opponent so Marines can close in. Typically, Marines target the opponent's limbs, but any portion of the body that is presented can become a target.

Vertical Slash Technique. The vertical slash follows a vertical line straight down through the target. To execute the vertical slash, Marines—

Thrust the right hand out and bring the weapon straight down on the opponent, continuing to drag the knife down through the opponent's body.

Maintain contact on the opponent's body with the blade of the knife.

Forward Slash Technique. The forward slash follows a straight line in a forehand stroke, across the target areas of either the neck (high diagonal slash) or abdominal region (low horizontal slash). To execute the forward slash, Marines-

Extend the right hand while simultaneously rotating the palm up until the knife blade makes contact with the opponent.

Snap or rotate the wrist through the slashing motion to maximize blade's contact with the opponent.

Drag the knife across the opponent's body, from right to left, in a forehand stroke. The movement ends with the forearm against the body and the knife at the left hip with its blade oriented toward the opponent.

Reverse Slash Technique. The reverse slash is a follow-up technique to a forward attack. It allows Marines both a secondary attack and the ability to resume the basic warrior stance. The reverse slash follows a straight line in a backhand stroke, across the target areas of either the neck (high diagonal slash) or abdominal region (low horizontal slash). To execute the reverse slash, Marines—

Extend the right hand while simultaneously rotating the palm down until the knife blade makes contact with the opponent.

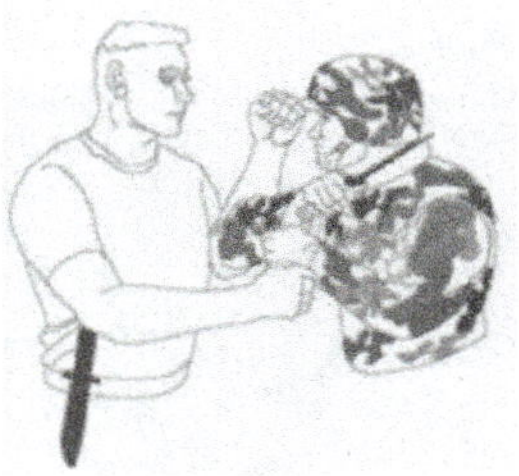

Snap or rotate the wrist through the slashing motion to maximize the blade's contact with the opponent.

Drag the knife across the opponent's body, from left to right, in a backhand stroke. Maintain contact on the opponent's body with the blade of the knife.

Thrusting Techniques

The primary objective of knife fighting is to insert the blade into an opponent to cause extensive damage and trauma. This is done with a thrusting technique. Thrusting techniques are more effective than slashing techniques because of the damage they can inflict. However, Marines use slashing techniques to close with the enemy so that they are closer to the opponent, which allows them to use the thrusting technique.

Vertical Thrust. The thrusting motion follows a vertical line straight up through the target (low into the abdomen region or high into the neck). To execute the vertical thrust, Marines—

Thrust the right hand toward the target, inserting the knife blade straight into the opponent.

Pull the knife out of the opponent.

Forward Thrust. The forward thrust follows a straight line straight into the opponent's neck (high thrust) or abdominal region (low thrust). To execute the forward thrust, Marines—

Thrust the right hand, palm down, toward the target, inserting the knife blade straight into the opponent.

Rotate the palm up once the knife is inserted to twist the blade.

Drop the right elbow and bring the knife to the opposite side of the opponent's body from where it was inserted. At the same time, rotate the hips and shoulders downward to bring body weight to bear on the attack.

Reverse Thrust. The reverse thrust is a follow-up technique to a forward attack. It allows Marines both a secondary attack and the ability to resume the basic warrior stance. The reverse thrust follows a horizontal line straight into the opponent's neck (high thrust) or abdominal region (low thrust). To execute the reverse thrust, Marines-

Bend the right arm, crossing the arm to the left side of the body.

Thrust the right hand, palm up, toward the target, inserting the knife blade straight into the opponent.

Rotate the palm down to twist the blade once the knife is inserted.

Bring the knife to the opposite side of the opponent's body from where it was inserted.

3. Weapons of Opportunity

During an unarmed close combat situation, Marines use their bodies as weapons, but they should be ready and able to use anything around them as a weapon. For example, Marines could throw sand or liquid in an opponent's eyes to temporarily impair his vision or smash the opponent's head with a rock or helmet. Marines must use whatever means are available and do whatever it takes to take control of the situation and to win, or they face the possibility of losing their lives. Some weapons of opportunity are discussed in the following subparagraphs.

Entrenching Tool

An entrenching tool (E-tool) is commonly carried by Marines. It can be an excellent weapon, especially when sharpened. Marines can use the E-tool to block, slash, and thrust at an opponent.

Tent Pole and Pins

Marines can use tent poles and pins to block, strike, or thrust at an opponent.

Web Belt

Marines can stretch a web belt between their hands to block attacks by an opponent.

Battlefield Debris

Marines can use debris on the battlefield (e.g., sticks, glass, a sharp piece of metal) to cut, slash, or stab an opponent. They can also use other types of debris such as shovel or ax handles, boards, metal pipes, or broken rifles to strike an opponent or apply a choke.

Helmet

A helmet can be used to strike an opponent on an unprotected area like the head and face. Grasp the rim of the helmet and thrust the arms forward, striking the opponent with the top of the helmet.

4. Fundamentals of Combative Stick

On the battlefield, Marines must be ready and able to use anything as a weapon. They must learn and be able to use techniques that can be employed with most weapons of opportunity. Among these techniques are combative stick techniques. Combative stick techniques can be used with a stick, a club, a broken rifle, an E-tool, or even a web belt.

Angles of Attack

There are six angles from which an attack with a hand-held weapon can be launched:

Vertical strike coming straight down on an opponent.
Forward diagonal strike coming in at a 45-degree angle to the opponent.
Reverse diagonal strike coming in at a 45-degree angle to the opponent.
Forward horizontal strike coming in parallel to the ground.
Reverse horizontal strike coming in parallel to the ground.
Forward thrust coming in a straight line to the opponent.

Grip

Grasp the stick about 2 inches from its base.

Stance

The basic warrior stance serves as the foundation for combative stick techniques. The left hand becomes a vertical shield that protects the ribs or the head and neck. Depending on how heavy the weapon is, it should be held at a level approximately shoulder height.

Movement

Movement during combative stick techniques is the same as it is for other close combat techniques. Marines can move anywhere within a 360-dcgrcc circle around the opponent. This allows accessibility to different target areas of the opponent's body and gains a tactical advantage.

Marines should avoid being directly in front of an opponent because the opponent can rely on his forward momentum to seize the tactical advantage. If Marines face an opponent, movement is made in a 45-dcgrcc angle to cither side of the opponent. This angle avoids an opponent's strike and places Marines in the best position to attack an opponent.

5. Combative Stick Techniques

Strikes

Strikes are intended to inflict as much damage on an opponent as possible. Striking techniques apply to a weapon of opportunity such as a stick, a tent pole, a club, a broken rifle, an E-tool, or a Pipe-

Vertical Strike. To execute the vertical strike, Marines—

Bend the right arm, extending the weapon over the back of the right shoulder.

Rotate the forearm straight down off the elbow to bring the weapon down on the opponent.

Rotate the hips and shoulders forcefully toward the opponent.

Follow through with the strike by allowing the weight of the weapon to go through the target area of the body.

Forward Strike. To execute a forward strike, Marines—

Step forward with the left foot in the direction of the strike.

Bend the right arm with the elbow extending out to the right and the weapon extended over the right shoulder.

Rotate the forearm to the right of the elbow to bring the weapon down onto the opponent. At the same time, forcefully rotate the hips and shoulders toward the opponent.

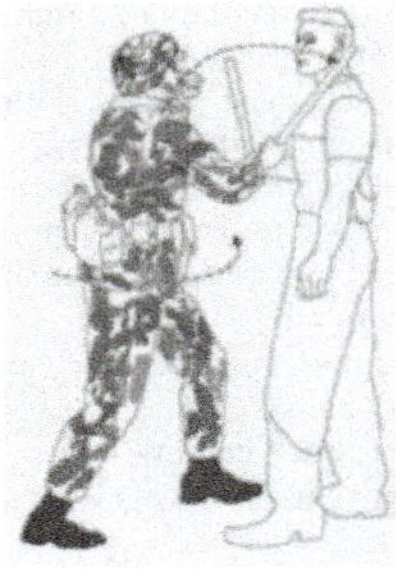

Follow through with the strike by allowing the weight of the weapon to go through the target area of the body.

Reverse Strike. The reverse strike is a follow-up technique to a forward strike. It allows Marines both a secondary attack and the ability to resume the basic warrior stance.To execute a reverse strike, Marines—

Step forward with the right foot in the direction of the strike.

Bend the right arm with the hand near the left shoulder. The weapon is extended over the left shoulder.

Rotate the forearm to the right of the elbow to bring the weapon down onto the opponent. At the same time, forcefully rotate the hips and shoulders toward the opponent.

Follow through with the strike by allowing the weight of the weapon to go through the target area of the body.

Forward Thrust. To execute the forward thrust, Marines—

Grasp the stick with the left hand, palm up, in a position where the stick can be controlled with two hands.

Lift the left leg and lunge forward off the ball of the right foot. At the same time, thrust the end of the weapon directly toward the opponent by thrusting both hands forward in a straight line.

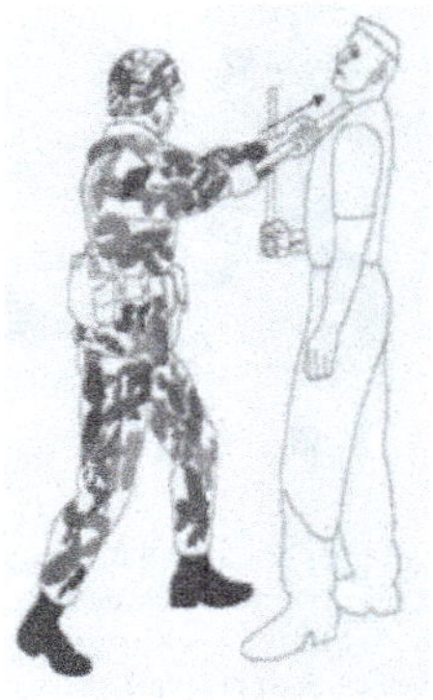

6. Blocking Techniques

A block is meant to deter or deflect an attack by an opponent. A block sets up Marines for a follow-on attack against the opponent. Blocks are executed by deflecting, rather than hitting or following through like a strike.

Blocks Against Unarmed Attacks

To block against an unarmed attack, Marines—

Step forward at a 45-degree angle with either the right or left foot. This moves the body out of the line of attack.

Raise the left arm and block the strike with the meaty portion of the forearm.

Employ the stick by one of two techniques:

Strike the opponent with the stick.
Use the stick to block on two points of contact. When the stick is used to block, it serves as an incidental strike. This technique is only used if Marines have closed with the opponent and are inside his strike.

Blocks Against Armed Attacks

Block for a Vertical Strike. To execute the block for a vertical strike, Marines—

Step forward with the left foot at a 45-degree angle to the left. This moves the body out of the line of attack.

Block on two points of contact to disperse the impact of the attack:

Block the opponent's stick by positioning the stick so it is perpendicular to the opponent' s stick. If the stick is not perpendicular to the opponent's stick, the stick can slide through and make contact on the Marine.
Block the opponent's wrist or forearm with the back of the left forearm.

Note: Use the stick to block the opponent's arm if closer to the opponent. It is the same movement as blocking with the arm, except the opponent's arm is blocked with both the stick and the arm.

Block for a Forward Strike. To execute the block for a forward strike, Marines-
Step forward with the left foot at a 45-degree angle to the left. This moves the body out of the line of attack and inside the opponent's strike.

Block on two points of contact to disperse the impact of the attack:

- Block the opponent's wrist or forearm with the meaty portion of the back of the left forearm.
- Strike the opponent's attacking biceps with the stick.

Block for a Reverse Strike. To execute the block for a reverse strike, Marines—

Step forward at a 45-degree angle with the right foot to the right. This moves the body out of the line of attack and inside the opponent's strike.

Block on two points of contact to disperse the impact of the attack:

Block the opponent's stick by positioning the stick so it is perpendicular to the opponent's stick.

Block the opponent's forearm with the meaty portion of the left forearm.

Note: If closer to the opponent, block his triceps with the back of the left forearm and strike his forearm with the stick.

7. Unarmed Against Hand-Held Weapons

If Marines are engaged against an opponent with a knife, a stick, or some other weapon of opportunity, they must establish and maintain an offensive mindset, not a defensive mindset. Their survival depends on it. They cannot afford to think about getting cut or hurt.

Angles of Attack

Before Marines can learn to block or counter an attack with a hand-held weapon (e.g., knife, stick), they must know from what angle the opponent is attacking. There arc six angles from which an opponent will typically attack with a hand-held weapon:

Vertical strike coming in straight down on the Marine.
Forward diagonal strike coming in at a 45-degree angle to the Marine.
Reverse diagonal strike coming in at a 45-degree angle to the Marine.
Forward horizontal strike coming in parallel to the ground.
Reverse horizontal strike coming in parallel to the ground.
Forward thrust coming in a straight line to the Marine.

Blocks

If the opponent has a hand-held weapon (e.g., knife or stick), Marines parry the opponent's hand or arm to block the attack.

Basic Block. To execute the basic block technique, Marines-

Step forward at a 45-degree angle to move out of the line of the attack. Always step in the direction of the strike.

Thrust the forearms forward, hands up, against the opponent's attacking arm. Contact is made on the opponent's arm with the backs of the forearms.

Block for a Vertical Strike. To execute the block against a vertical strike, Marines—

Step forward with the left foot at a 45-degree angle to the left to move out of the line of attack.

Thrust the forearms forward, hands up, against the outside of the opponent's attacking arm.

Block for a Forward Strike. To execute the block against a forward diagonal or forward horizontal strike, Marines-

Step forward with the left fool inside the opponent's attacking arm.

Block the attack with both arms bent so the forearms make contact with the opponent's biceps and forearm.

Block for a Reverse Strike. To execute the block against a reverse diagonal or reverse horizontal strike, Marines—

Step forward with the right foot to the outside of the opponent's attacking arm.

Block the attack with both arms bent so the forearms make contact with the opponent's triceps and forearm.

Block for a Forward Thrust. To execute the block against a forward thrust, Marines—

Bend at the waist, move the hips backward, and jump backward with both feet to move away from the attack. This action is known as "hollowing out."

Hollow out and block the attack with the arms bent and hands together on top of the attacking arm.

Overlap the hands slightly so one thumb is on top of the other hand's index finger. The other thumb should be under the other hand's index finger.

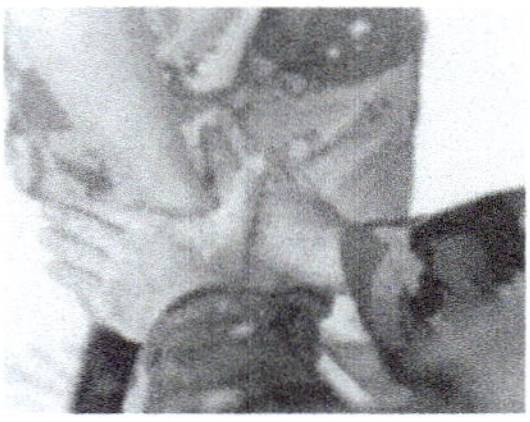

8. Counters to Hand-Held Weapon Attacks

Principles of Counters

A counter is used to control the situation to regain the tactical advantage and end the fight. Regardless of the type of weapon or angle of attack, the following principles apply to countering the attack with a hand-held weapon:

Move out of the line of attack. Movement is executed in a 45-degree angle forward to the left or right.

- Block the attack.

Note: The first two actions are taken simultaneously.

- Take control of the weapon by controlling the hand or arm that is holding the weapon. Never attempt to grab the opponent's weapon.
- Execute the appropriate follow-up to end the fight; e.g., strikes, joint manipulations, throws, or takedowns (see chap. 8). Marines should continue to attack the opponent until the fight ends.

Counter Techniques

There are two techniques that can be used to counter any armed attack: forward armbar counter and reverse armbar counter. These techniques can be used to counter a vertical attack, a forward diagonal strike, or a forward horizontal strike. With minor variations, the same techniques arc used to counter reverse strikes. A third technique, the bent armbar counter, is used to counter a vertical attack.

Forward Armbar Counter. To execute the forward armbar counter to an attack coming from a forward strike, Marines-

Step forward with the left foot inside the opponent' s attacking arm.

Block the attack with both arms bent so the forearms make contact with the opponent's biceps and forearm.

Slide the left arm over the opponent's forearm and wrap the arm tightly around his arm, trapping the opponent's attacking arm between the biceps and torso.

Place the right hand on the opponent's shoulder or upper arm to further control his arm and to effect an armbar.

Execute an armbar and continue to exert steady pressure against the arm.

Reverse Armbar Counter. To execute the reverse armbar counter to an attack coming from a forward strike, Marines-

Step forward with the left foot inside the opponent' s attacking arm.

Block the attack with both arms bent so the forearms make contact with the opponent's biceps and forearm.

Control the opponent's arm with the left hand and pivot to the right so the back is against the opponent's side. Immediately slide the right arm over the opponent's biceps and wrap the arm tightly around his arm, trapping the opponent's attacking arm between the biceps and torso.

Grasp the opponent's wrist with the left hand and twist his thumb away from the body.

Take the opponent to the ground with an armbar takedown if he does not drop the weapon from the wristlock.

Bent Armbar Counter. This counter is particularly effective against a vertical attack. To execute the bent armbar counter, Marines—

Step forward with the left foot inside the opponent's attacking arm.

Block the attack with both arms bent so the forearms make contact with the opponent's biceps and forearm.

Grasp the opponent's forearm with the left hand. At the same time, slide the right arm underneath the opponent's triceps and grasp the opponent's forearm or wrist with the right hand.

Apply downward pressure with the hands against the opponent's forearm to off-balance the opponent.

CHAPTER 4

STRIKES

This chapter describes all techniques for a right-lmnded person. However, all techniques can be executed from either side.

In drawings, the Marine is depicted in woodland camouflage utilities; the opponent is depicted without camouflage. In photographs, the Marine is depicted in woodland camouflage utilities; the opponent is depicted in desert camouflage utilities.

Strikes are unarmed individual hitting techniques. Strikes use the hands, elbows, knees, feet, and, in some instances, other parts of the body as personal weapons. Marines must know how to execute strikes effectively. They must also know how to counter strikes from an opponent.

1. Principles of Punches

Muscle Relaxation

Muscle relaxation is crucial when executing punches. The natural tendency in a fight is to tense up, which results in rapid fatigue and decreased power generation. Marines who remain relaxed during a close combat situation generate greater speed, which results in greater generation of power. Relaxing the forearms generates speed and improves reaction time. At the point of impact, Marines clench the fist to cause damage to the opponent and avoid injury to the wrist and hand.

Weight Transfer

Weight transfer is necessary to generate power in a punch. Marines accomplish this by-

Rotating their hips and shoulders into the attack.

Moving their body mass straight forward or backward in a straight line.

Dropping their body weight into an opponent. Body mass can be transferred into an attack from high to low or from low to high.

Rapid Retraction

When Marines deliver a punch, rapid retraction of the fist is important. Once the hand has made contact with the target, Marines quickly return to the basic warrior stance. Rapid retraction-

Returns the hand and arm to the protection afforded by the basic warrior stance.

Prevents the opponent from being able to grab the hand or arm.

Permits the hand and arm to be "chambered" or "re-cocked" in preparation for delivering a subsequent punch.

Telegraphing

Telegraphing a strike occurs when body movements inform the opponent of the intention to launch an attack. Staying relaxed helps to reduce telegraphing.

Often, an untrained fighter telegraphs his intention to attack by drawing his hand back in view of his opponent, changing facial expression, tensing neck muscles, or twitching. These movements, however small, immediately indicate an attack is about to be delivered. If the opponent is a trained fighter, he may be able to evade or counter the attack. If the opponent is an untrained fighter, he may still be able to minimize the effect of an attack.

2. Punches

Punches may be thrown during any hand-to-hand confrontation. Most people resort to punching because it is a natural reaction to a threat. The purpose of a punch is to stun the opponent or to set him up for a follow-up finishing technique. However, punches should only be executed to the soft tissue areas of an opponent. A correctly delivered punch maximizes the damage to an opponent while minimizing the risk of injury to Marines.

Basic Fist

Punches are executed using the basic fist. To make the basic fist, the fingers are curled naturally into the palm of the hand and the thumb is placed across the index and middle fingers. Do not clench the fist until movement has begun. This reduces muscular tension in the forearm and increases speed and reaction time. Just before impact, Marines exert muscular tension on the hand and forearm to maximize damage to the opponent and reduce their chances of injury. Contact should be made with the knuckles of the index and middle fingers.

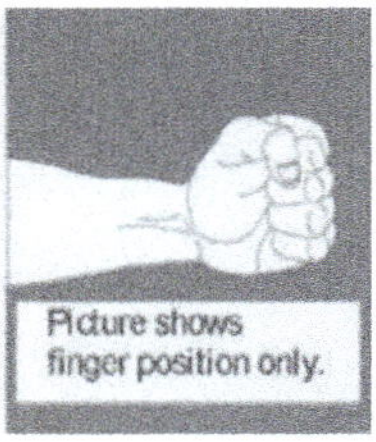

When striking with the basic fist, Marines must keep the hand straight, or in line, with the wrist to avoid injury to the wrist.

Lead Hand Punch

The lead hand punch is a snapping straight punch executed by the forward or lead hand. It is a fast punch designed to keep the opponent away and to set up an attack. A lead hand punch conceals movement and allows Marines to get close to the opponent. Lead hand punches should strike soft tissue areas, if possible. To execute the lead hand punch, Marines-

Snap the lead hand out to nearly full extension, while rotating the palm to the ground.

Contact the opponent with the first two knuckles of the fist.

Retract the hand immediately, resuming the basic warrior stance.

Rear Hand Punch

The rear hand punch is a snapping punch executed by the rear (right) hand. It is a power punch designed to inflict maximum damage on the opponent. Its power comes from pushing off the rear leg and rotating the hips and shoulders. To execute the rear hand punch, Marines—

Rotate the hips and shoulders forcefully toward the opponent and thrust the rear hand straight out, rotating the palm down, to nearly full extension.

Shift body weight to the lead foot while pushing off on the ball of the rear foot.

Contact the opponent with the first two knuckles of the fist.

Retract the hand immediately.

Uppercut

The uppercut is a powerful punch originating below the opponent's line of vision. It is executed in an upward motion traveling up the centerline of the opponent's body. It is delivered in close and usually follows a preparatory strike that leaves the target area unprotected. When delivered to the chin or jaw, the uppercut can render an opponent unconscious, cause extensive damage to the neck, or sever the tongue. To execute the uppercut, Marines—

Bend the arms, rotating the palm inboard. The distance the arms bend depends on how close the opponent is.

Rotate the hips and shoulders forcefully toward the opponent, thrusting the fist straight up toward the opponent's chin or jaw.

Contact the opponent with the first two knuckles of the fist.

Retract the hand immediately.

Hook

The hook is a powerful punch that is executed close in and is usually preceded by a preparatory strike. To execute the hook, Marines—

Thrust the right arm in a hooking motion toward the opponent, keeping the elbow bent while forcefully rotating the right shoulder and hip toward the opponent.

Contact the opponent with the first two knuckles of the fist. Continue rotating the shoulder and hip, following through with the fist to the target.

Retract the hand immediately.

3. Strikes with the Upper Body

Strikes stun the opponent or set him up for a follow-up finishing technique. The hands, forearms, and elbows are individual weapons of the arms that can be used to execute strikes including the hammer fist, chin jab, knife hand, eye gouge, and elbow strikes. These strikes provide a variety of techniques that can be used in any type of close combat situation.

Principles of Execution

There are several principles of execution that ensure a strike's effectiveness.

Generating Power. Weight transfer is necessary to generate maximum power in a punch. Marines accomplish this by-

- Rotating the hips and shoulders into the attack.
- Moving the body mass straight forward or backward in a straight line.
- Dropping body weight into an opponent.
- Body mass can be transferred into an attack from high to low or from low to high.

Muscular Tension. The arms are relaxed until the moment of impact. At the point of impact, Marines apply muscular tension in the hand and forearm to maximize damage to the opponent and to avoid injury to the hand. The arms are relaxed until the moment of impact.

Hit and Stick and Follow-Through. A strike should be delivered so that the weapon (e.g., hand, elbow) hits and remains on the impact site (target) and follows through the target. This technique inflicts maximum damage on the opponent.

Strikes with the arms are executed with "heavy hands"; i.e., the strike is executed by driving through with the strike to allow the weight of the hand to go through the target area of the body. Contact with the opponent should be made with the arm slightly bent, and the arm extends as it moves through the target. This technique allows Marines to deliver strikes effectively without executing full force.

Movement. Movement puts Marines in the proper position for launching an attack against an opponent as well as providing protection.

Movement is initiated from the basic warrior stance and ends by resuming the basic warrior stance. Strikes can be performed with either the left or right arm depending upon—

The angle of attack.
The position of the opponent.
The opponent's available, vulnerable target areas.

Target Areas of the Body

For each strike, there are target areas of the body that, when struck, maximize damage to an opponent. Strikes use gross motor skills as opposed to fine motor skills. The target areas of the body are just that—areas. Pinpoint accuracy on a specific nerve is not needed for the strike to be effective.

Hammer Fist

Striking with the hammer fist concentrates power in a small part of the hand, which, when transferred to the target, can have a devastating effect. The striking surface of the hammer fist is the meaty portion of the hand below the little finger. To execute the hammer fist strike, Marines-

Make a fist and bend the arm at approximately a 45- to 90-degree angle. At the same time, rotate the right hip and right shoulder backward.

Thrust the fist forward onto the opponent while rotating the right hip and shoulder forward.

Rotate the wrist so the hammer fist makes contact on the opponent.

Chin Jab

The chin jab can immediately render an opponent unconscious and cause extensive damage to the neck and spine. The striking surface is the heel of the palm of the hand. To execute the chin jab, Marines—

Bend the right wrist back at a 90-degree angle with the palm facing the opponent and the fingers pointing up.

Keep the right arm bent and close to the body. Extend the hand into a concave position with the fingers slightly spread apart.

Step forward with the left foot toward the opponent, keeping the feet approximately shoulder-width apart and the knees bent. This is done to close with the opponent.

Keep the right arm bent and close to the side. Thrust the palm of the hand directly up under the opponent's chin. At the same time, rotate the right hip forward to drive the body weight into the attack to increase the power of the strike. The attack should travel up the centerline of the opponent's chest to his chin.

Knife Hand

The knife hand is one of the most versatile and devastating strikes. The striking surface is the cutting edge of the hand, which is the meaty portion of the hand below the little finger extending to the top of the wrist. The striking surface is narrow, allowing strikes on the neck between the opponent's body armor and helmet. The knife hand strike is executed from one of three angles: outside, inside, and vertical.

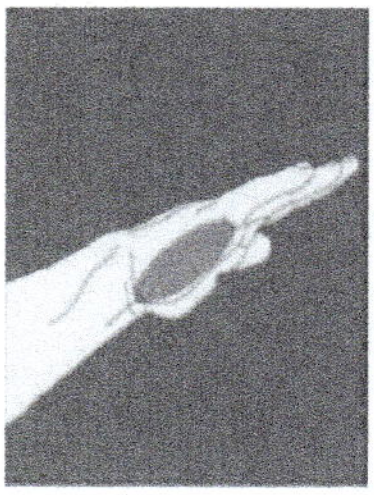

Outside Knife Hand. To execute the outside knife hand strike, Marines—

Execute a knife hand by extending and joining the fingers of the right hand and placing the thumb next to the forefinger (like saluting).

Retract the right hand. At the same time, rotate the right hip and right shoulder backward.

Thrust the knife hand forward (horizontally) onto the opponent while rotating the right hip and shoulder forward.

Inside Knife Hand. To execute the inside knife hand strike, Marines—

Execute a knife hand.

Bring the right hand over the left shoulder. At the same time, rotate the right shoulder forward and the left hip forward.

Thrust the knife hand forward (horizontally) onto the opponent while rotating the right hip and shoulder forward and the left shoulder backward.

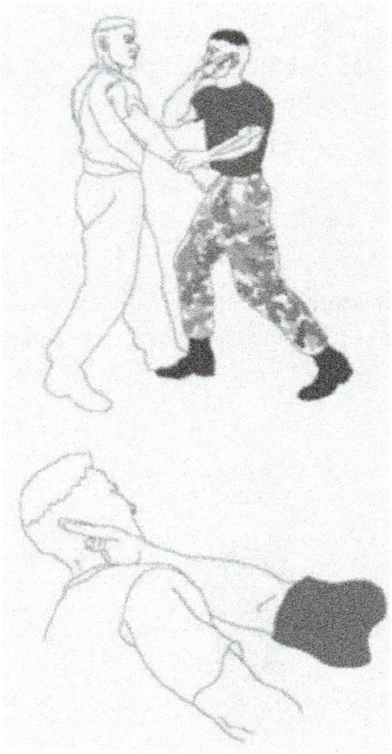

Vertical Knife Hand. When thrown vertically, the knife hand strike comes straight down in a straight line.

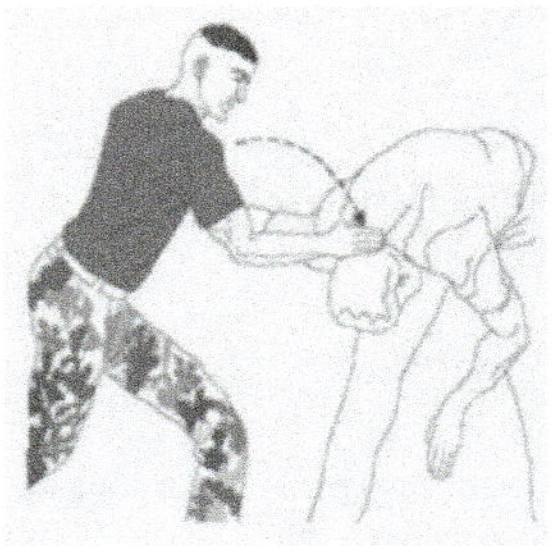

Eye Gouge

The eye gouge is used to attack an opponent's eyes, blinding him so follow-up strikes can be executed. The striking surface is the tips of the fingers or thumb. To execute the eye gouge, Marines—

Extend the right hand with the fingers slightly spread apart to allow entry into the eye sockets.

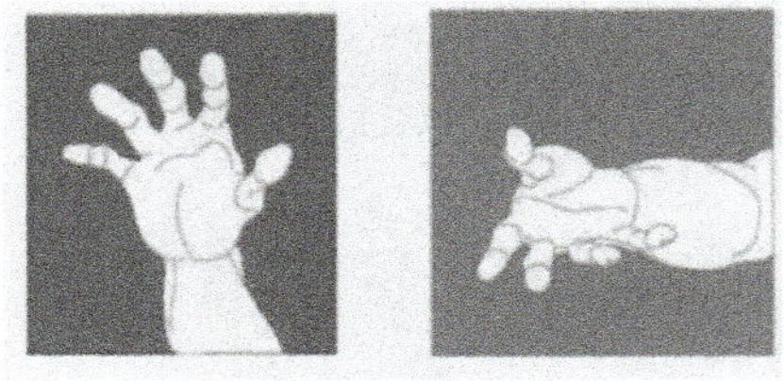

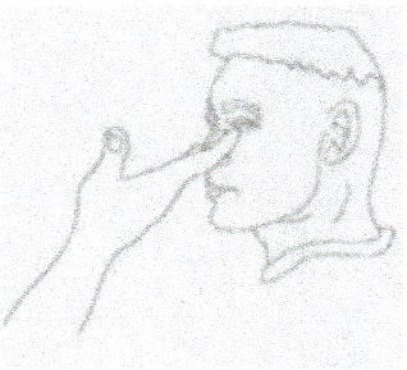

Place the palm of the hand either toward the ground or toward the sky and thrust the right hand forward into the opponent's eyes.

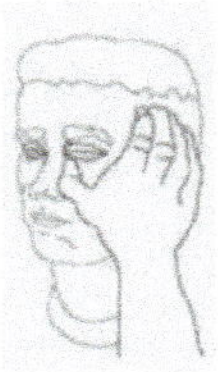

Thrust the hand forward at the opponent's nose level so the fingers or thumb slide naturally into the grooves of the opponent's eye sockets.

Elbow Strikes

The elbow is a powerful weapon that can be used in several different ways to attack virtually any part of an opponent's body. Elbow strikes can be performed either vertically (upward or downward) or horizontally (forward or reverse). The striking surface is 2 inches above or below the point of the elbow, depending upon the angle of attack, the opponent's attack angle, and the position of the opponent.

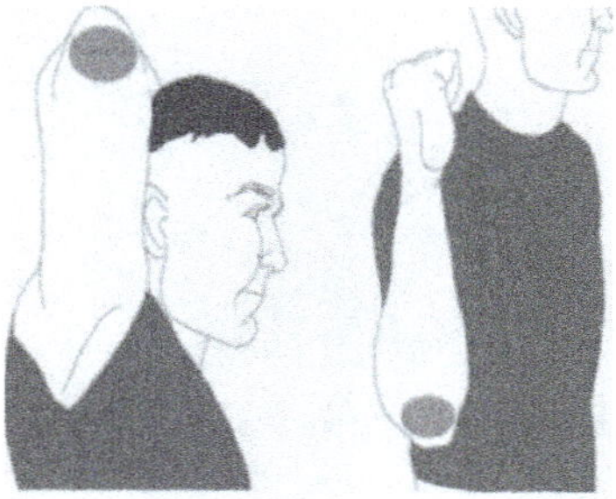

Above Elbow Below Elbow

Vertical Elbow Strike (Up). To execute an upward vertical elbow strike, Marines-

Bend the right elbow, keeping the fist close to the body. The fist is at shoulder level and the elbow is next to the torso.

Thrust the elbow vertically upward toward the opponent while rotating the right shoulder and hip forward to generate additional power.

Contact the opponent with the right forearm 2 inches above the point of the elbow.

Vertical Elbow Strike (Down). To execute a downward vertical elbow strike, Marines—

Bend the right elbow, keeping the fist close to the body. The fist is on the shoulder and the elbow is raised well above the shoulder.

Thrust the elbow vertically downward toward the opponent while dropping body weight into the attack to generate additional power.

Contact the opponent with the right triceps 2 inches above the point of the elbow.

Horizontal Elbow Strike (Forward). To execute the forward horizontal elbow strike, Marines-

Tuck the right fist near the chest with the palm heel facing the ground.

Thrust the right elbow horizontally forward toward the opponent. The forearm is parallel to the ground.

Rotate the right shoulder and hip forward to generate additional power.

Contact the opponent with the right forearm 2 inches below the point of the elbow.

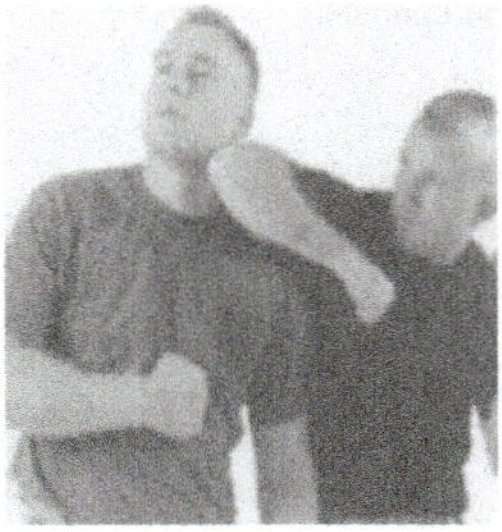

Horizontal Elbow Strike (Rear). To execute the rear horizontal elbow strike, Marines-

Tuck the right fist near the left shoulder with the palm heel facing the ground. At the same time, rotate the right shoulder forward and the left hip forward.

Thrust the right elbow horizontally rearward toward the opponent. The forearm is parallel to the ground and the hand moves toward the direction of the attack.

Rotate the right hip back and the right shoulder backward to generate additional power.

Contact the opponent with the right triceps 2 inches above the point of the elbow.

4. Strikes with the Lower Body

The legs are the body's most powerful weapons because they use the largest muscle groups to generate a strike. Legs are also less prone to injury. The feet are the preferred choice for striking because they are protected by boots. Marines use their feet, heels, and knees to execute kicks, knee strikes, and stomps.

Kicks

The purpose of a kick is to stop an opponent's attack or to create an opening in his defense in order to launch an attack. Kicks can be performed with the left (lead) leg or the right (rear) leg. Kicks with the rear leg have greater power because the hips are rotated into the attack. However, the rear leg is further away from the opponent so a strike with the rear leg will not contact the opponent as quickly as a strike with the lead leg.

Front Kick. The front kick is executed when the opponent is in front of the Marine. The front kick, delivered with either the rear or lead leg, is effective for striking below the waist. Attempting to kick higher results in diminished balance and provides the enemy with a greater opportunity to grab the leg or foot. The striking surfaces are the toe of the boot or the bootlaces. To execute the front kick, Marines-

Raise the left knee waist high, pivot the hips into the attack, and thrust the left foot forward toward the opponent.

Contact the opponent with the toe of the left boot or bootlaces.

Return to the basic warrior stance.

Side Kick. The side kick, delivered with the lead leg, is effective for striking the knees. The side kick is executed when the opponent is to the side of the Marine. The striking surface is the outside cutting edge of the boot near the heel. To execute the side kick, Marines—

Raise the right knee waist high and rotate the right hip forward.

Thrust the right foot to the right side toward the opponent, turning the foot at a 90-degree angle to maximize the striking surface on the opponent.

Contact the opponent with the cutting edge of the right boot.

Return to the basic warrior stance.

Knee Strikes

Knee strikes are excellent weapons during the close range of close combat fighting. The knee strike is generally delivered in close.

Vertical Knee Strike. The striking surface is the thigh, slightly above the knee. To execute the vertical knee strike, Marines-

Raise the right knee and drive it up forcefully into the opponent. Power is generated by thrusting the leg upward.

Contact the opponent 2 inches above the right knee.

Horizontal Knee Strike. The horizontal knee strike is executed with the leg generally parallel with the ground while rotating the hips to generate power. It is often delivered to the peroneal nerve. The striking surface is the front of the leg, slightly above or below the knee. To execute the horizontal knee strike, Marines-

Raise the right knee, rotate the right hip forward while pivoting on the left foot, and drive the knee horizontally into the opponent.

Contact the opponent 2 inches above the right knee.

Stomps

Stomps are delivered with the feet, usually when the opponent is down. Remember, when the opponent is down, Marines take whatever target is available.

Vertical Stomp. The vertical stomp allows Marines to remain upright and balanced, to rapidly deliver multiple blows with either foot, and to quickly and accurately attack the target. It is the preferred stomp. The striking surface is the flat bottom of the boot or the cutting edge of the heel. To execute the vertical stomp, Marines—

Raise the right knee above the waist with the right leg bent at approximately a 90-degree angle.

Drive the flat bottom of the right boot or the cutting edge of the right heel down onto the opponent forcefully. At the same time, bend the left knee slightly to drop the body weight into the strike.

Ax Stomp. The striking surface of the ax stomp is the cutting edge of the heel. To execute the ax stomp, Marines-

Raise the right heel above the waist, keeping the right leg straight.

Drive the cutting edge of the right heel down onto the opponent forcefully. At the same time, bend the left knee slightly to drop the body weight into the strike.

5. Counters to Strikes

In a close combat situation, an opponent will attempt to strike Marines with punches and kicks. When an opponent uses a strike, Marines must first avoid the strike. This is accomplished by moving quickly and blocking. Next, Marines must get into an offensive position. This allows Marines to use offensive strikes to attack the opponent. Regardless of the strike, the counter to a strike requires Marines to move, block, and strike.

Move

The first step in countering a strike is to move out of the way of the strike's impact. Movement removes Marines from the opponent's intended strike point and positions Marines to attack. Movement is executed at approximately a 45-degree angle to the front or rear. Movement is always initiated from the basic warrior stance. Return to the basic warrior stance with the toe of the lead foot pointing toward the opponent once the movement is complete.

Block

Different blocks are executed based on the strike. These will be covered by individual counters.

Strike

Any of the upper body or lower body strikes can be executed as a follow-on attack or part of the counter to an opponent's strike. The follow-on strike is determined by the angle to the opponent, the position of the opponent, and the opponent's available vulnerable target areas.

Counters to Punches

Counter to a Lead Hand Punch. This counter is used when the opponent throws a lead hand punch. To execute the counter to the lead hand punch, Marines-

Step forward and to the left at approximately a 45-degree angle, moving in to the outside of the opponent's attacking arm.

Raise the left arm and block or deflect the opponent's lead hand with the palm of the hand or the meaty portion of the forearm.

Hit and stick by leaving the left arm against the opponent's right arm while stepping forward and to the left at approximately a 45-degree angle to close with the opponent.

Execute a finishing technique, such as a strike or a kick, to the opponent's exposed target areas.

Counter to a Rear Hand Punch. This counter is used when the opponent throws a rear hand punch. To execute the counter to the rear hand punch, Marines—

Step forward and to the left at approximately a 45-degree angle, moving in to the outside of the opponent's attacking arm.

Raise the left arm and block or deflect the opponent' s rear hand with the palm of the hand or the meaty portion of the forearm.

Hit and stick by leaving the left arm against the opponent's right arm while stepping forward and to the right at approximately a 45-degree angle to close with the opponent.

Execute a finishing technique, such as a strike or a kick, to the opponent's exposed target areas.

Counters to Kicks

Counter to a Front Kick (Left or Lead Leg). This counter is used when the opponent executes a front kick with his left leg. To execute the counter to a front kick, Marines-

Step forward and to the right at approximately a 45-degree angle, moving in to the outside of the opponent's striking leg.

Lower the left arm and block or deflect the opponent' s leg with the palm of the hand or the meaty portion of the forearm.

Hit and stick by leaving the left arm against the opponent's leg while stepping forward and to the left at approximately a 45-degree angle to close with the opponent.

Execute a finishing technique, such as a strike or a kick, to the opponent's exposed target areas.

Counter to a Front Kick (Right or Rear Leg). This counter is used when the opponent executes a front kick with his right leg. To execute the counter to a front kick, Marines—

Step forward and to the left at approximately a 45-degree angle, moving in to the outside of the opponent's striking leg.

Lower the left arm and block or deflect the opponent' s leg with the palm of the hand or the meaty portion of the forearm.

Hit and stick by leaving the left arm against the opponent's leg while stepping forward and to the right at approximately a 45-degree angle to close with the opponent.

CHAPTER 5

THROWS

This chapter describes all techniques for a right-handed person. However, all techniques can be executed from either side.

The Marine is depicted in camouflage utilities. The opponent is depicted without camouflage.

Marines use throwing techniques to maintain the tactical advantage and to throw the opponent to the ground during close combat. Throws apply the principles of balance, leverage, timing, and body position to upset an opponent's balance and to gain control by forcing the opponent to the ground. Throwing techniques are effective because they are size- and gender-neutral, and they rely on the momentum and power generated by the opponent rather than the strength or size of the Marine. Marines also execute a throw as a devastating attack against an opponent, possibly causing unconsciousness or broken limbs. When Marines execute throws, they must maintain balance and, simultaneously, prevent the opponent from countering a throw or escaping after being forced to the ground.

1. Turning Throw

Marines use a turning throw to take the opponent to the ground while they remain standing. A turning throw can also be executed from a stationary position. It is particularly effective if Marines and the opponent are wearing gear. To execute the turning throw, Marines-

Grasp the opponent's right wrist with the left hand.

Step forward with the right foot, place it against the outside of the opponent's right foot, and pivot so the back of the heel is next to the middle of the opponent's foot.

Hook the opponent's right arm with the right arm and pinch his arm between the biceps and forearm, touching the opponent with the body.

Pull the opponent's wrist downward, keeping it close to the body.

Pivot to the left on the ball of the foot and continue pulling downward on the opponent's wrist while rotating the wrist outward to off-balance him.

2. Hip Throw

Marines use a hip throw to take the opponent to the ground while they remain standing. A hip throw is particularly effective if the opponent is moving forward or pushing. Marines use the opponent' s forward momentum to execute the hip throw. To execute the hip throw, Marines-

Grasp the opponent's right wrist with the left hand.

Step forward with the right foot and place it against the outside of the opponent's right foot.

The back of the heel should be next to the opponent' s foot.

Step back with the left foot and rotate on the ball of the foot. The back of the heel is next to the opponent' s toe. The knees are bent.

Rotate at the waist and hook the right hand around the back of the opponent's body (anywhere from his waist to his head). The side and hip should be against the opponent.

Rotate the hip against the opponent. The hips must be lower than the opponent's.

Use the right hand to pull the opponent up on the hip to maximize contact.

Pull the opponent's arm across the body and, at the same time, lift the opponent off the ground slightly by bending at the waist, straightening the legs, and rotating the body to the left.

3. Leg Sweep

Marines use a leg sweep to take the opponent to the ground while they remain standing. A leg sweep is particularly effective if the opponent is already off-balanced and moving backward or pulling on the Marine. To execute the leg sweep, Marines—

Grab the opponent's right wrist with the left hand and grab the opponent' s left shoulder with the right hand.

Note: Grab the opponent's clothing or gear if his wrist and shoulder cannot be grabbed.

Step forward with the left foot and place it on the outside of the opponent's right foot.

Pull the opponent's wrist downward, close to the body, and push his shoulder backward to off-balance him.

Raise the right knee no higher than the waist.

Kick the foot past the opponent's right leg.

Use the heel of the boot to make contact with the opponent's calf (anywhere from the top of the calf down to the Achilles tendon or on the inside of the calf).

Sweep the opponent to the ground.

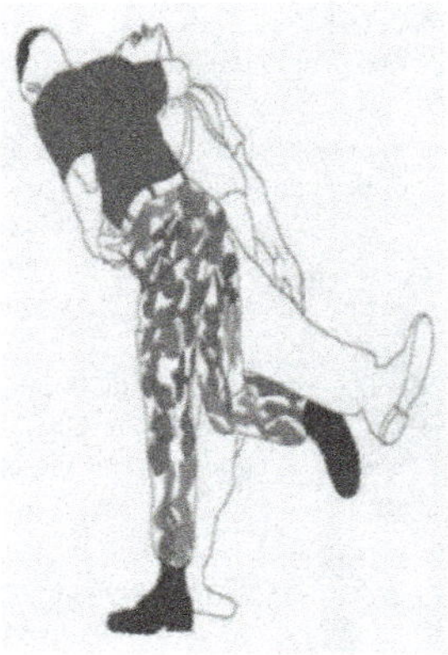

CHAPTER 6

CHOKES AND HOLDS

This chapter describes all techniques for a right-handed person. However, all techniques can be executed from either side.

The Marine is depicted in camouflage utilities. The opponent is depicted without camouflage.

WARNING

During training, never execute a choke at full force or full speed and never hold a choke for more than 5 seconds.

When Marines correctly perform a choke, they render an opponent unconscious in as little as 8 to 13 seconds. Chokes are easily performed regardless of size or gender. Marines must know how to apply chokes and how to counter a choke or a hold executed by an opponent.

1. Types of Chokes

There are two types of chokes: an air choke and a blood choke. An air choke closes off the airway to the lungs, thereby preventing oxygen from reaching the heart. A blood choke cuts off the blood flow to the brain. Both types can result in unconsciousness and eventual death for an opponent.

Air Choke

An air choke is performed on the opponent's windpipe (or trachea), cutting off the air to the lungs and heart. If Marines execute the air choke properly, the opponent loses consciousness within 2 to 3 minutes. Due to the length of time it takes to immobilize the opponent, air chokes are not recommended.

Blood Choke

A blood choke is performed on the opponent's carotid artery, which carries oxygen-enriched blood from the heart to the brain. The carotid artery is located on both sides of the neck. If Marines execute a blood choke properly, the opponent will lose consciousness within 8 to 13 seconds. The blood choke is the preferred choke because its intended effect (i.e., the opponent losing consciousness) can be executed quickly, ending the fight.

2. Chokes

Front Choke

Marines execute a front choke when they are facing the opponent. Marines use the opponent's lapels or collar to execute a front choke. To execute the front choke, Marines-

Grab the opponent's right lapel with the right hand, making certain that the knuckles or the back of the hand are against the opponent's carotid artery on the right side of his neck.

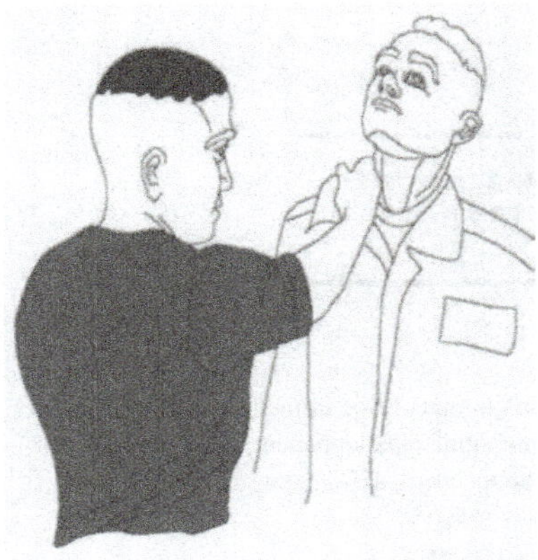

Keep the right hand against the opponent's neck, reach under the opponent's right arm with the left hand, grab the opponent's left lapel, and form an X with the wrists.

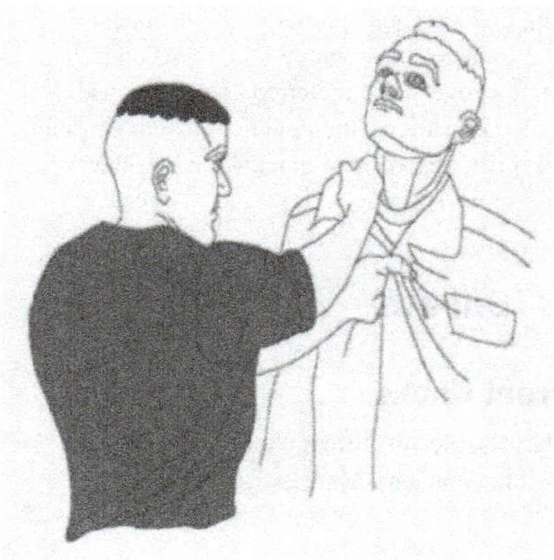

Press the right hand against the opponent's carotid artery.

Pull his lapel to the left with the left hand.

Side Choke

Marines execute a side choke when they are facing the opponent. The side choke is particularly effective when deflecting a punch thrown by an opponent. To execute the side choke, Marines—

Use the left hand to parry the opponent's arm inboard (to the inside of the opponent's reach).

Bring the right arm underneath the opponent's arm and up around the front of his neck.

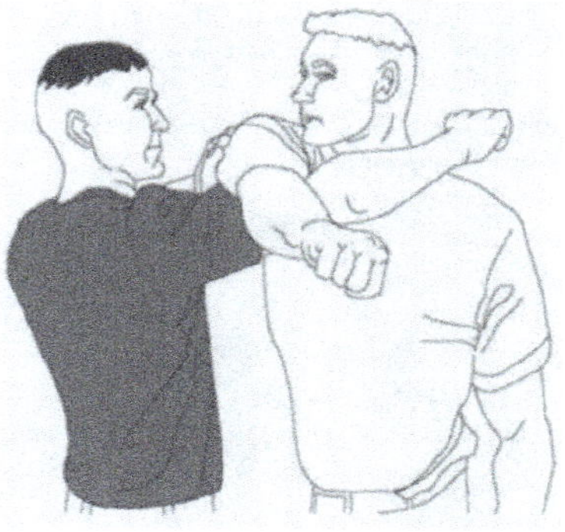

Extend the fingers, place the back of the forearm against the opponent's neck just below his ear, and press the carotid artery on his neck.

Note: If Marines are unable to place their fingers against the opponent's neck, they may use the back of the thumb or wrist.

Reach, with the left hand, around the back of the opponent's neck and clasp the hands together.

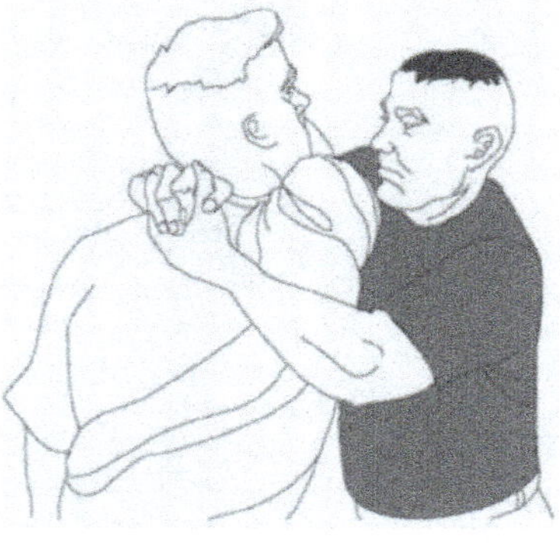

Pull the opponent toward the chest by pulling the clasped hands toward the chest.

Exert pressure on the side of the opponent's neck with the forearm.

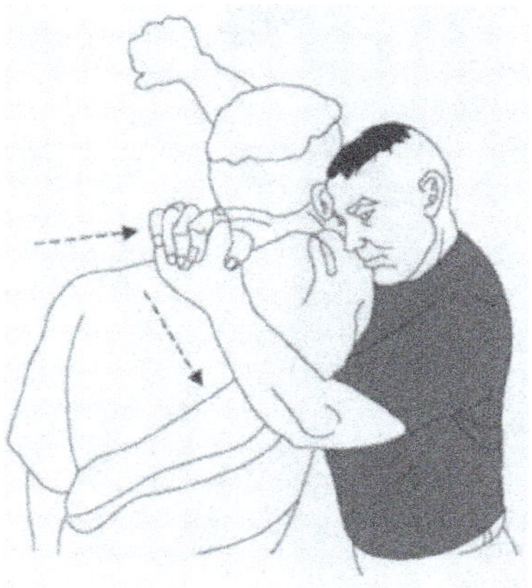

Rear Choke

Marines execute a rear choke when they are behind the opponent, the opponent is on the ground, or when they are taking the opponent to the ground. To execute the rear choke, Marines—

Reach, with the right arm, over the opponent's right shoulder and hook the bend of the arm around his neck.

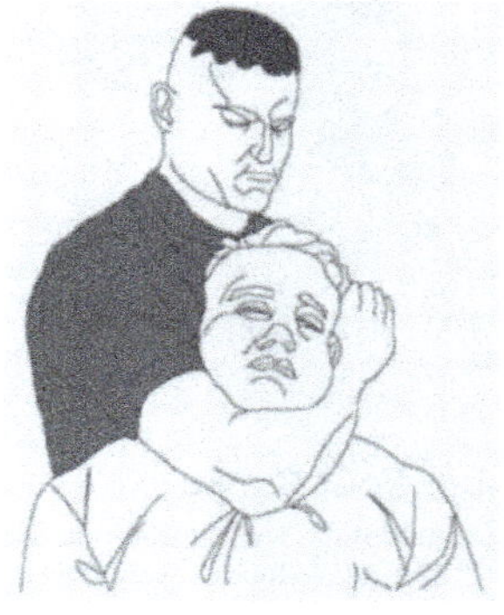

Clasp both hands together.

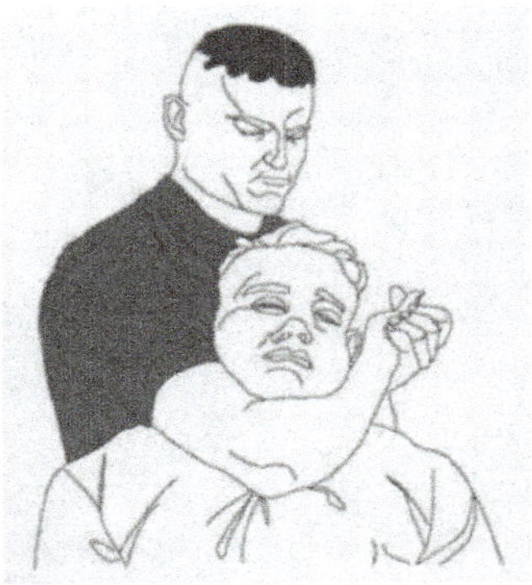

Exert pressure with the biceps and forearms on both sides of the opponent's neck on his carotid arteries.

Maintain pressure with the biceps and forearms on both sides of the neck and draw the opponent closer by drawing the right arm in.

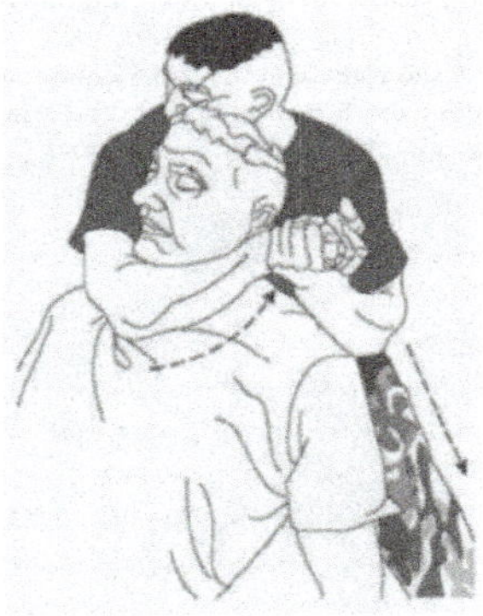

Figure-Four Choke

A variation of the rear choke is the figure-four choke. The figure-four choke allows Marines to gain more leverage than the rear choke. If the rear choke cannot be secured, the figure-four variation is employed to increase the pressure of the choke on the opponent. To execute the figure-four variation of the rear choke, Marines—

Apply a rear choke. The Marine's body should be against the opponent's body.

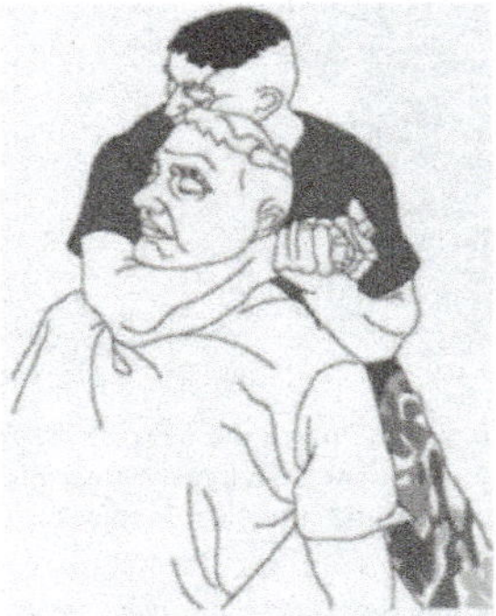

Grasp the left biceps with the right hand and place the left hand against the back of the opponent's head.

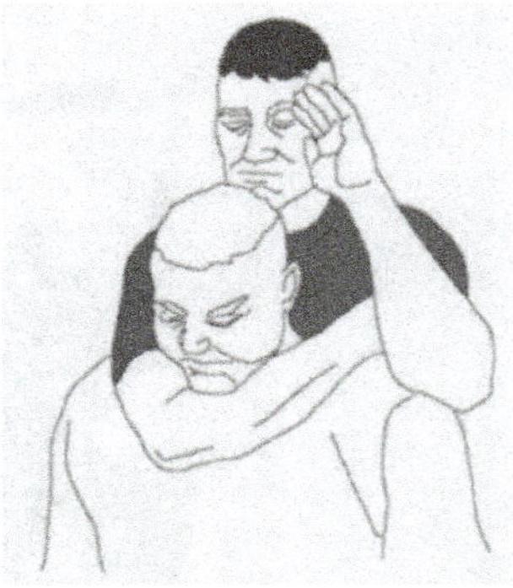

Push the opponent's head forward and down with the left hand.

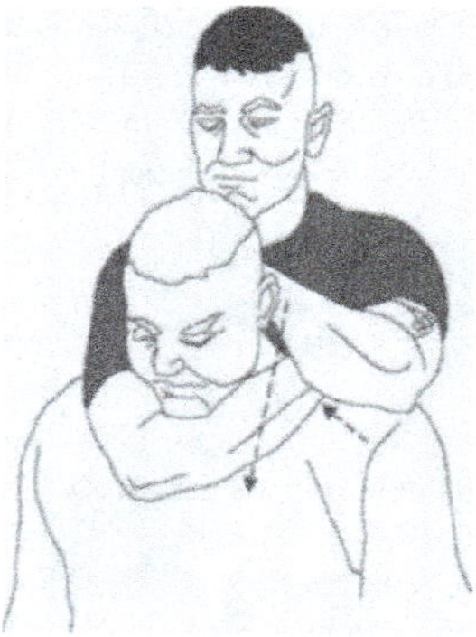

Draw the right arm in, maintaining pressure with the bicep and forearm on both sides of the opponent's neck.

3. Counters to Chokes and Holds

During a close combat situation, an opponent may apply a choke or hold on a Marine. If the opponent correctly applies a choke, a Marine quickly loses consciousness. If a choke is not executed properly, it often results in a hold, typically a bear hug or a headlock. A hold allows the opponent to control a Marine and removes the Marine's ability to attack. It is important for Marines to extract themselves from chokes and holds, regain the tactical advantage, and counter with strikes.

Although a choke causes unconsciousness in 8 to 13 seconds for a blood choke and 2 to 3 minutes for an air choke, the first movement in any counter to a choke is to clear the airway. Marines use softening techniques to loosen an opponent's grip and to clear their airway. Softening techniques are particularly effective if Marines lack the physical strength of their opponent. These techniques include groin strikes, eye gouges, foot stomps, etc. Softening techniques are not offensive; rather, they are used to loosen an opponent's hold.

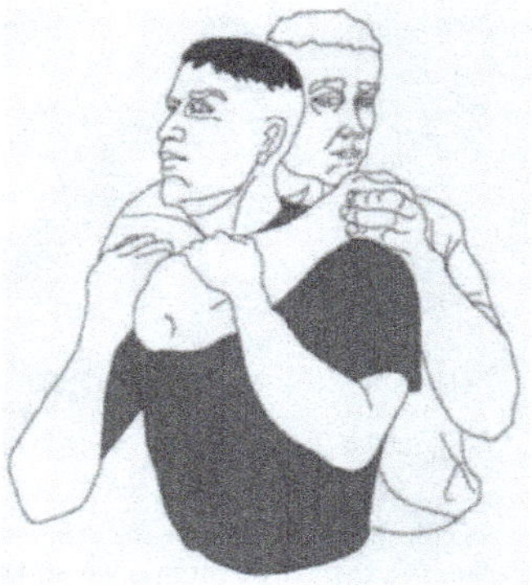

The second movement is to tuck the chin. Once the airway is clear, Marines tuck their chins to prevent the opponent from reapplying the choke.

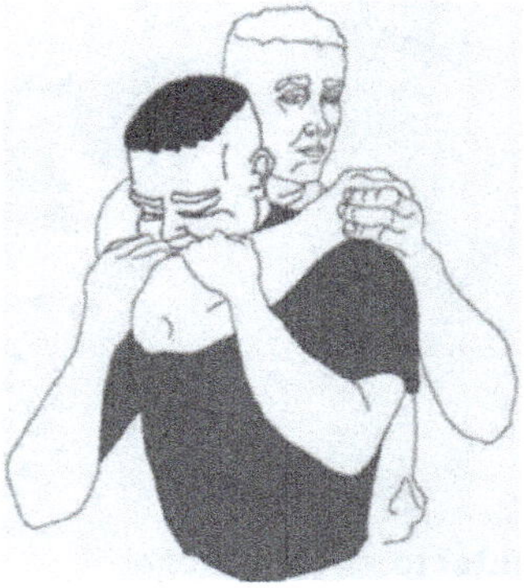

Counter to a Front Choke

Marines use a counter to a front choke when the opponent approaches from the front and uses both hands to choke a Marine around the throat. To execute the counter to the front choke, Marines—

Grasp the opponent's right forearm (where the elbow bends) with the left hand and apply downward pressure on the opponent's radial nerve with the fingers.

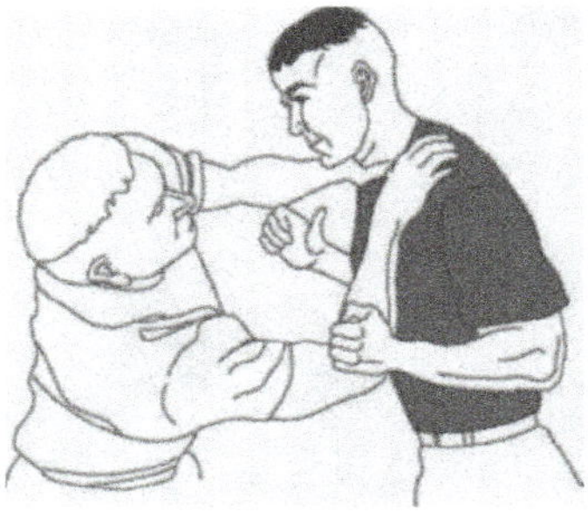

Execute a chin jab to the opponent's chin with the right hand. To generate power into the strike, bring the left foot to the outside of the opponent's right foot and rotate the hips into the strike.

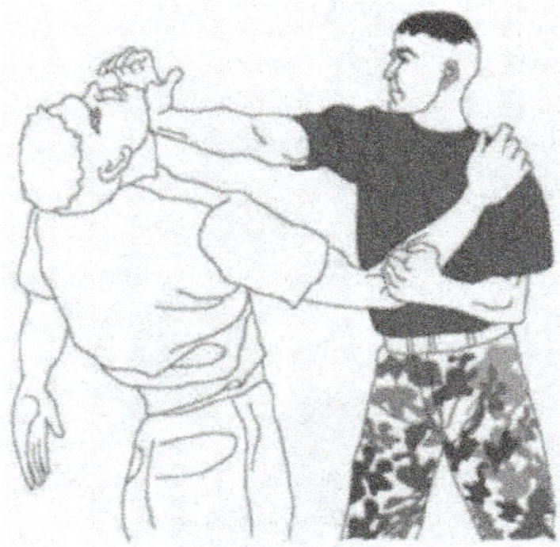

Counter to a Rear Choke

Marines execute a counter to a rear choke when the opponent approaches from the rear and puts his right arm around a Marine's throat. To execute the counter to the rear choke, Marines-

Grasp the opponent's forearm (at the radial nerve) and bicep with both hands and pull down just enough to clear the airway. Once the airway is clear, tuck the chin to protect the airway and to prevent the opponent from reapplying the choke.

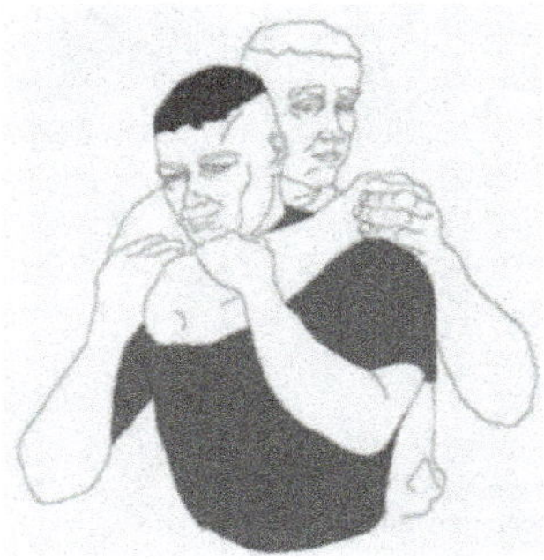

Step behind the opponent's right leg with the left foot, keeping both legs bent (almost in a squatting position).

Strike and drive the left elbow into the opponent's torso while rotating the hips and pivoting to the left, throwing the opponent back and to the ground.

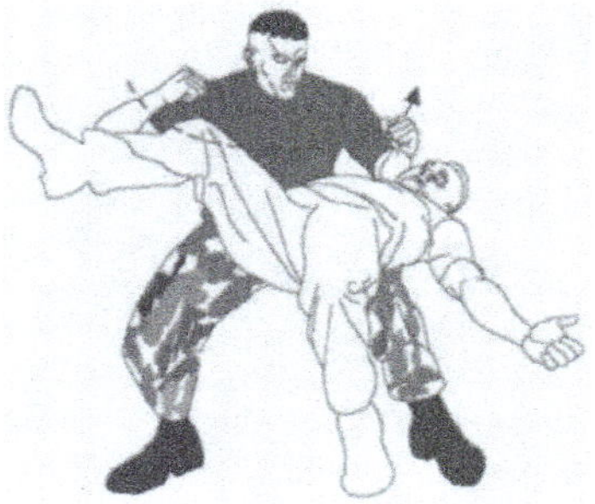

Counter to a Front Headlock

Marines use a counter to a front headlock when the opponent approaches from the front and puts his right arm around the Marine's neck, bends the Marine forward, and locks the Marine's head against his hip. To execute the counter to a front headlock, Marines—

Grasp the opponent's wrist and forearm with both hands and pull down to clear the airway. Maintain control of the opponent's wrist throughout the move. Once the airway is clear, tuck the chin to protect the airway and to prevent the opponent from reapplying the choke.

Move the right hand and arm across the opponent's torso.

Step forward and to the left with the left foot at a 45-degree angle.

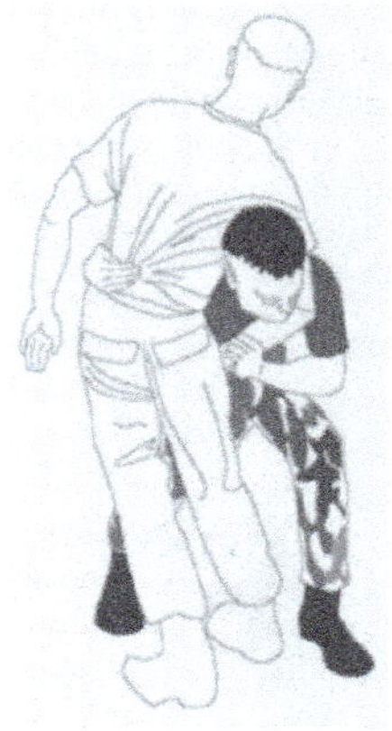

Execute a sweep with the right foot against the opponent's right leg. At the same time, push against the opponent's chest with the right arm and shoulder to generate power in the sweep.

Counter to a Rear Headlock

Marines use a counter to a rear headlock when the opponent approaches from the rear and puts his right arm around the Marine's neck, bends the Marine forward, and locks the Marine's head against his hip. To execute the counter to the rear headlock, Marines-

Grasp the opponent's wrist and forearm with the right hand and pull down to clear the airway. Once the airway is clear, tuck the chin to protect the airway and to prevent the opponent from re-applying the choke.

Reach over the opponent's right shoulder with the left arm.

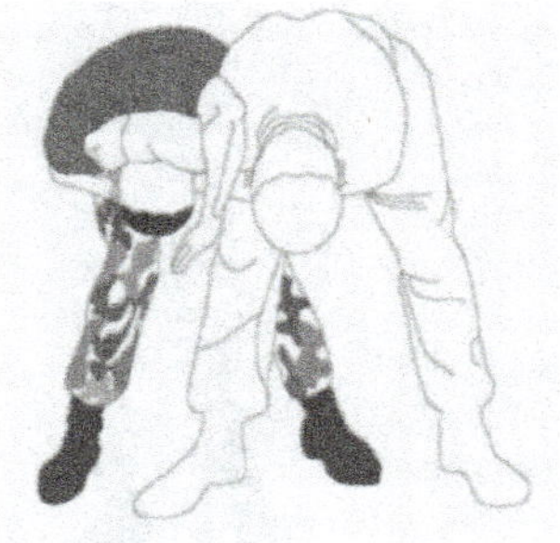

Grab any part of the opponent's face (chin, nose, eyes) and pull back while rising to a standing position.

Execute, with the right hand, a hammer fist strike to the opponent's exposed throat.

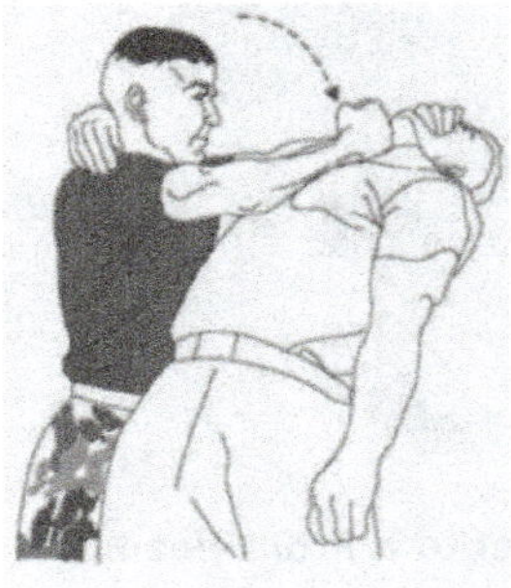

Counter to a Front Bear Hug

Marines execute a counter to a front bear hug when the opponent approaches from the front and puts both of his arms around the Marine's body, trapping the Marine's arms to the sides. To execute the counter to a front bear hug, Marines—

Step forward and to the left with the left foot at a 45-degree angle to the outside of the opponent's right leg, keeping the left leg bent.

Grasp the opponent's torso or arms to gain balance and to assist in throwing the opponent. It may be helpful to hook the opponent's right arm with the left arm.

Drive the right arm and shoulder forward and, at the same time, bring the right leg forward and sweep the opponent's right leg, bringing him to the ground.

Counter to a Rear Bear Hug

Marines use a counter to a rear bear hug when the opponent approaches from the rear and puts both of his arms around the Marine's body, trapping the Marine's arms to the sides. To execute the counter to a rear bear hug, Marines—

Step behind the opponent's right leg with the left foot, keeping both legs bent (almost in a squatting position). The left side of the body should be against the opponent's.

Pivot the hip, turning the body to the left and throwing the opponent back over the bent leg.

CHAPTER 7

GROUND FIGHTING

This chapter describes all techniques for a right-handed person. However, all techniques can be executed from either side.

In drawings, the Marine is depicted in woodland camouflage utilities; the opponent is depicted witlwut camouflage. In photographs, the Marine is depicted in woodland camouflage utilities; the opponent is depicted in desert camouflage utilities.

Marines should avoid being on the ground during a close combat situation because the battlefield may be covered with debris and there is an increased risk of injury. However, many close combat situations involve fighting on the ground. The priority in a ground fight is for Marines to get back on their feet as quickly as possible. In any ground fighting scenario, Marines will usually end up in one of four positions with the opponent. The offensive positions, in which Marines have a tactical advantage, are the guard and mount. The defensive positions, which are used as counters when the opponent has the tactical advantage, are the counter to the guard and the counter to the mount. Marines can also employ chokes during ground fighting to quickly end a fight.

1. Offensive Ground Fighting

Guard Position

Marines execute a guard position when the opponent is on top and the legs are wrapped around the opponent's legs. If the opponent is on top, he may try to choke the Marine, but the Marine still has the tactical advantage because the Marine is in a position to control the situation. To execute the guard position technique, Marines—

Trap the opponent's hands on the chest by crossing the hands on the chest so the forearms are resting on the opponent's forearms. Apply pressure with the elbows or forearms.

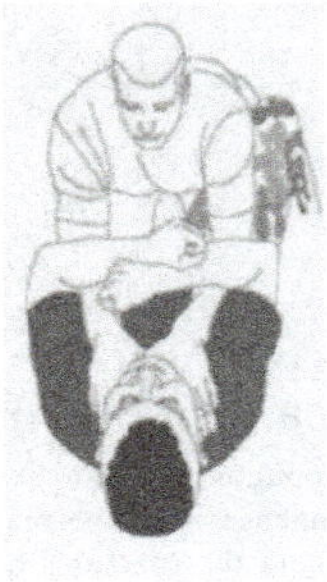

Strike the outside of the opponent's thigh with the cutting edge of the right heel. This causes the opponent to jerk to that side.

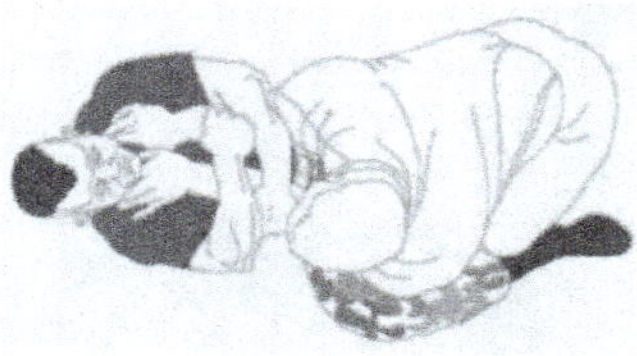

Move the head quickly to the left and swivel the hips to the right. At the same time, bring up both of the legs. Both legs are on the right side of the opponent's body.

Bring the right leg down, hooking the opponent's neck and head. Exert downward pressure to roll him over on his back. Grasp and maintain control of the opponent's left arm. Upon completion of the move, the Marine is sitting up with legs bent over the opponent while maintaining control of his arm.

Keep legs and knees bent. Maintain pressure against the opponent's neck with the back of the right foot and against his side with the left foot underneath his armpit. Squeeze the knees together, locking the opponent's arm.

Pull the opponent's arm straight up and fall back sharply, pulling his arm to the side in the direction of his little finger.

Return to standing.

Mount Position

Marines execute the mount if the opponent is lying on his back on the ground and the Marine is on top with legs wrapped around the opponent's body. This position is an offensive position because Marines are in a better position to control the opponent and to execute ground fighting techniques. To execute the mount position technique, Marines—

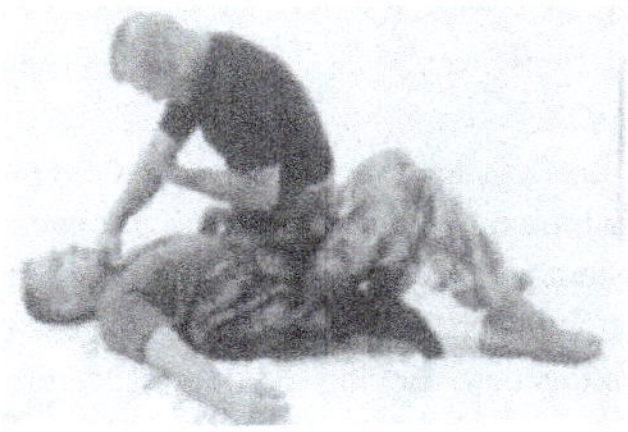

Grab the opponent's wrists or forearms with the hands. Hold them tightly against the torso.

Maintain control of the opponent's arms with the left arm and apply pressure to the opponent's brachial plexus (tie in) with the right hand.

Bring the left foot close to the opponent's right armpit and swing the right leg across the opponent' s head.

Keep the legs and knees bent. Maintain pressure against the opponent's neck with the back of the right foot and against his side with the left foot which is underneath his armpit. Squeeze the knees together, locking the opponent's arm.

Pull the opponent's arm straight up and fall back sharply, pulling his arm to the side in the direction of his little finger.

Return to standing.

2. Defensive Ground Fighting

Counter to the Guard

Marines use this technique if the opponent is lying on his back on the ground and the Marine is kneeling on the ground between the opponent's legs. To execute the counter to the guard position, Marines—

Strike the opponent's femoral nerve, located on the inside of the thigh, with the elbows. This forces the opponent to separate his legs.

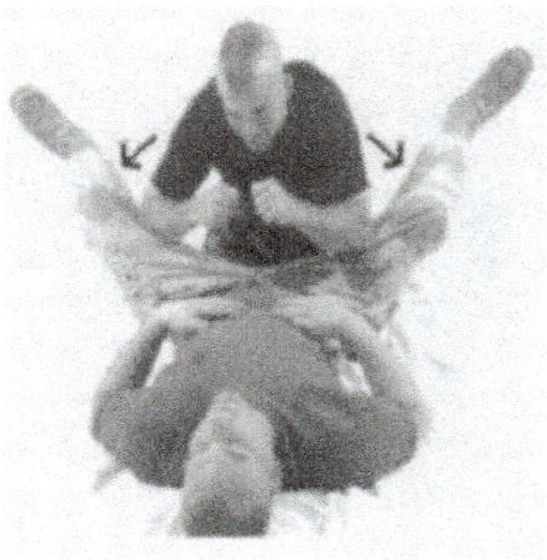

Strike the opponent's groin with the right fist.

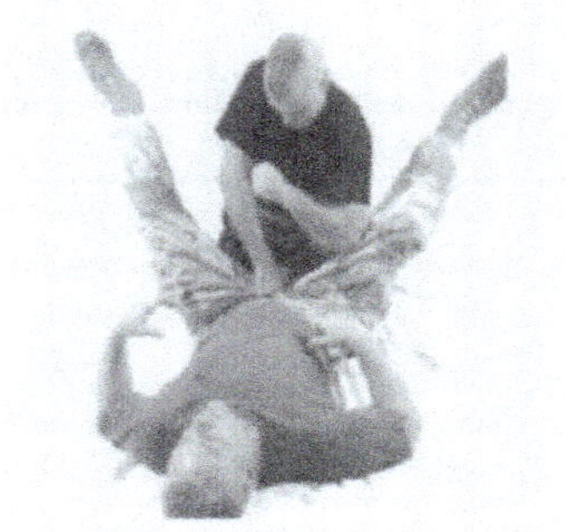

Hook the left arm underneath the opponent's right knee from the inside. Throw the opponent's leg over the head.

Duck the head quickly and move to the left.

Return to standing.

Counter to the Mount

Marines use this technique when they are lying on their back on the ground, the opponent is mounted on top of the Marine, and the opponent's legs are wrapped around the Marine. The opponent has the tactical advantage. To execute the counter to the mount position, Marines—

Grab the opponent's gear or clothing on his upper torso and pull him down close.

Use the right arm to hook the opponent's left arm, from the inside around the outside, above his elbow. With the right foot, hook the opponent's left leg or ankle.

Draw the elbow in to bend the opponent's elbow, bringing him down close. The arm must be hooked above the opponent's elbow in order to bend it.

Strike the opponent's side with the left hand.

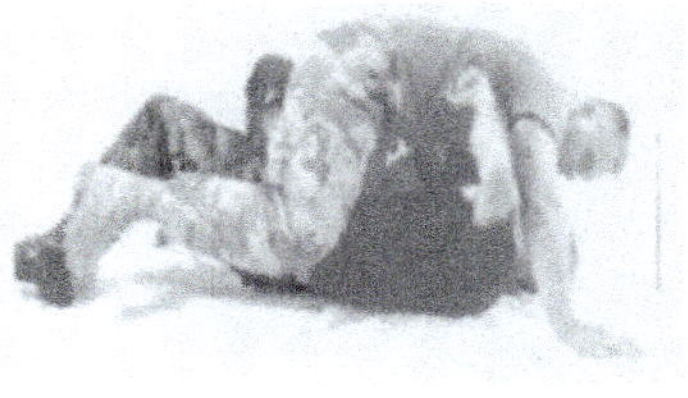

Push the opponent over and roll him off to the right side.

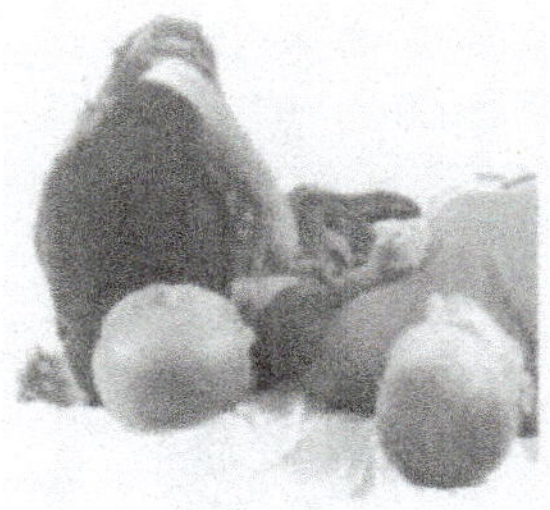

Maintain control of the opponent's hooked arm and move to a kneeling position.

Move to a standing position while maintaining pressure on the opponent's arm and using the knee to apply pressure against the opponent's elbow.

3. Ground Fighting Chokes

WARNING

During training, never execute a choke at full force or full speed. Never hold a choke for more than 5 seconds.

The priority in ground fighting is for Marines to get back on their feet as quickly as possible. Sometimes, Marines can quickly end a ground fight by executing a choke on the opponent. When performed correctly, a choke can render an opponent unconscious in as little as 8 to 13 seconds. Chokes are easily performed regardless of size or gender. The chokes performed during ground fighting are the same as those performed while standing (see chap. 6).

Ground Fighting Front Choke

The ground fighting front choke is a blood choke performed most effectively from the mount position. The front choke employs the opponent's lapels or collar to execute the choke. To execute the ground fighting front choke, Marines—

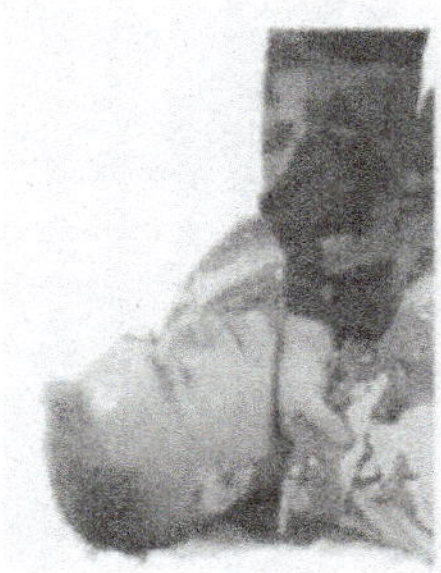

Grab the opponent's right lapel with the right hand, making certain that the knuckles or the back of the hand are against the carotid artery on the right side of the opponent' s neck.

Keep the right hand pressed against the opponent's carotid artery, reach under the right arm with the left hand, and grab the opponent's left lapel, forming an X with the wrists.

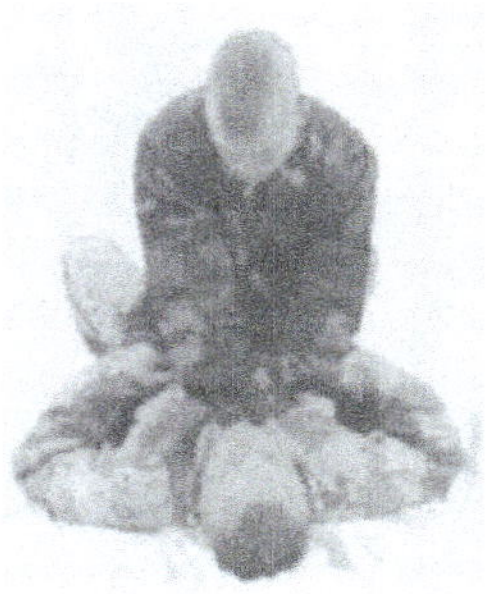

Keep the right hand pressed against the opponent's carotid artery and pull the opponent's left lapel to the left with the left hand.

Ground Fighting Side Choke

The ground fighting side choke is a blood choke performed from the mount position. The ground fighting side choke is particularly effective when the opponent raises his arms and places them on the Marine's chest or throat. To execute the ground fighting side choke, Marines-

Use the left hand to parry the opponent's right arm inboard (to the inside of the opponent's reach). Bring the right arm underneath the opponent's arm and up around the front of his neck.

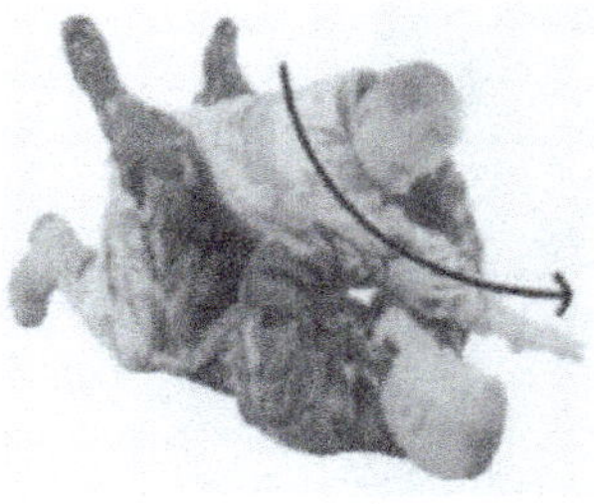

Keep the fingers extended, place the back of the forearm against the opponent's neck just below his ear and press the carotid artery on his neck. Depending on the Marine's or the opponent's body configuration, the back of the thumb or wrist may press against the opponent's neck.

Use the left hand to reach around the back of the opponent's neck and clasp the hands together.

Pull the opponent toward the chest. Use the forearm to exert pressure on the side of his neck. This is done by pulling the clasped hands toward the chest.

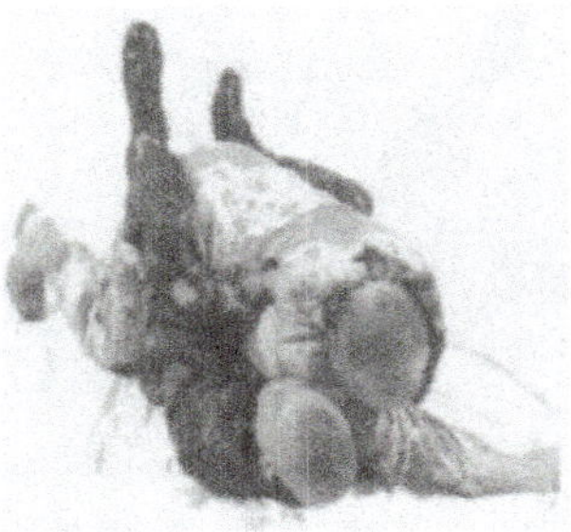

Ground Fighting Rear Choke

The ground fighting rear choke is a blood choke performed when Marines are behind the opponent. To execute the ground fighting rear choke, Marines—

Wrap the legs around the opponent with the left foot against the inside of his left thigh and the right foot against the inside of his right thigh.

Use the right arm to reach over the opponent's right shoulder and hook the bend of the arm around his neck.

Use the left hand to clasp both hands together.

Exert pressure with the biceps and forearms on both sides of the opponent's neck. This applies pressure to the opponent's carotid arteries. While maintaining pressure with the biceps and forearms, draw the opponent closer by drawing the right arm in. To increase the effectiveness of the choke, lean backward by arching the back and pulling the opponent back. At the same time, push the feet against the opponent's thighs.

Figure-Four Choke

A variation of the ground fighting rear choke is the figure-four choke. The figure-four choke allows Marines to gain more leverage on the rear choke. If the rear choke cannot be secured, the figure-four variation may be applied to increase the pressure of the choke on the opponent. To execute the figure-four variation of the rear choke, Marines—

Apply the ground fighting rear choke. The chest should be against the opponent's back.

Grasp the left biceps with the right hand and place the left hand against the back of the opponent's head.

Use the left hand to push the opponent's head forward and down.

Draw the right arm in while maintaining pressure with the biceps and forearm on both sides of the opponent's neck.

Increase the effectiveness of the choke by arching backward and pushing the feet against the opponent's thighs. Continue to pull the opponent back with the right arm while exerting pressure forward with the left hand against the opponent's head.

CHAPTER 8

NONLETHAL TECHNIQUES

This chapter describes all techniques for a riglit-handed person. However, all techniques can be executed from either side.

In drawings, the Marine is depicted in woodland camouflage utilities; the opponent is depicted without camouflage. In photographs, the Marine is depicted in woodland camouflage utilities; the opponent is depicted in desert camouflage utilities.

The Marine Corps' involvement in military operations other than war—e.g., humanitarian, peacekeeping, or evacuation missions—has greatly increased. These missions require skills that span the spectrum of conflict and support operations within a continuum of force. But the Corps' day-to-day existence also demands a responsible use of force. Nonlethal techniques are among the skills Marines use to apply a responsible use of force.

1. Unarmed Restraints and Manipulation

Marines operate within a continuum of force, particularly in support of peacekeeping or humanitarian types of missions. In these situations, Marines must act responsibly to handle a situation without resorting to deadly force. Unarmed restraints and manipulation techniques including joint manipulation, come-alongs, and takedowns can be used to control a subject without resorting to deadly force. Marines must train to become proficient in nonlethal techniques and to respond in a responsible manner. These techniques are referred to as compliance techniques, and they are applied in the third level in the continuum of force.

WARNING

During training, never apply the techniques for unarmed restraints and manipulation at full force or full speed. Use a slow and steady pressure to avoid injury.

Compliance Techniques

Compliance techniques are unarmed restraint and manipulation techniques used to physically force a subject or opponent to comply. Compliance can be achieved through the close combat techniques of-

Pain compliance using joint manipulation and pressure points. (Pain compliance is the initiation of pain to get compliance on the part of the subject.)

Come-along holds.

Principles of Joint Manipulation

Joint manipulation is used to initiate pain compliance and gain control of a subject. It involves the application of pressure on thejoints (elbow, wrist, shoulder, knee, ankle, and fingers). Pressure is applied in two ways:

In the direction in which the joint will not bend. For example, joints such as the knees

and elbows only bend in one direction and when pressure is applied in the opposite direction, pain compliance can be achieved. Beyond the point where the joint stops naturally in its range of movement (i.e., it no longer bends).

Since each joint has a breaking point, Marines should apply slow steady pressure only until pain compliance is reached. Continued pressure will break the joint and may escalate the violence of the situation.

Joint manipulation also uses the principle of off-balancing. A subject can be better controlled when he is knocked off balance.

Wristlocks

A wristlock is a joint manipulation that can be applied in a number of ways to achieve pain compliance. The wrist rotates in a number of directions and will bend in a single direction until its movement stops naturally. In a wristlock, pressure is exerted beyond that point by bending or twisting the joint. A wristlock is executed when an opponent tries to grab Marines or is successful in grabbing Marines or their equipment. A wristlock can also be performed by Marines if they wish to initiate control of an opponent.

Basic Wristlock. A basic wristlock is executed when Marines grab the opponent's left hand with the right hand. To execute the basic wristlock, Marines—

Use the right hand to grab the opponent's left hand by placing the thumb on the back of the opponent's hand so that the Marine's knuckles are facing to the left.

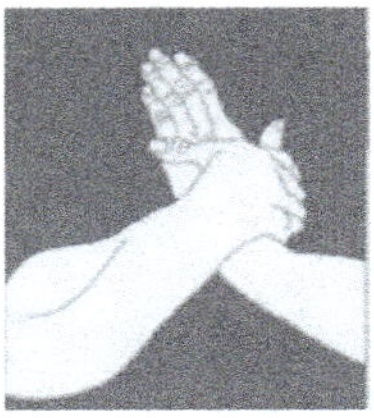

Hook the fingers across the fleshy part of the opponent's palm below the thumb. The fingers are used to anchor the hand so leverage can be applied to twist and bend the joint.

Exert downward pressure with the thumb to bend the opponent'sjoint. Rotate the opponent's hand to the right to twist the joint.

Use the left hand to further control the opponent.

Step in to the opponent to keep the opponent's hand in close to the body to control him and provide more leverage on the wristlock.

Note: When executing the basic wristlock with the left hand, the Marine grabs the opponent's hand so that the Marine's knuckles are facing to the right, and then rotates and twists the opponent's hand to the left.

Reverse Wristlock. A reverse wristlock is executed when Marines grab the opponent's right hand with the right hand. To execute the reverse wristlock, Marines—

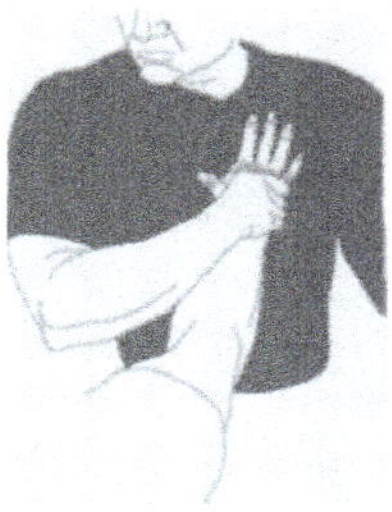

Place the right palm on the back of the opponent's right hand and wrap the fingers across the fleshy part of his palm below his little finger.

Twist the opponent's hand to the right while stepping in to place his hand against the chest. Apply downward pressure on the opponent's hand against the chest. Leave the opponent's hand on the chest to fully control the subject and to gain leverage.

Lean forward to use body weight to add additional pressure to the joint.

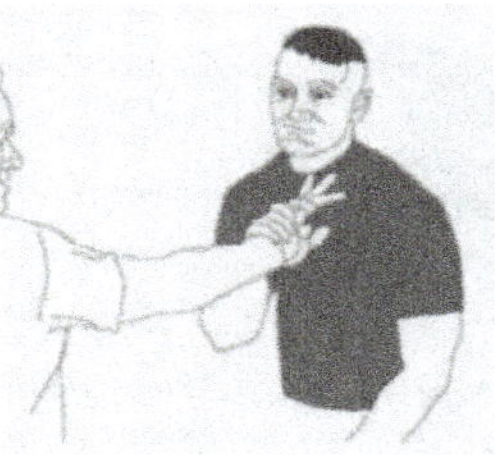

Use the left hand to further control the opponent.

Two-Handed Wristlock. Both hands can be used in the wristlock to maximize the leverage and pressure needed to bend and twist the joint. To execute the two-handed wristlock, Marines—

Place both thumbs on the back of the opponent's hand, thumbs crossed.

Hook the fingers of both hands around the fleshy part of the opponent's palm on both sides of his hand.

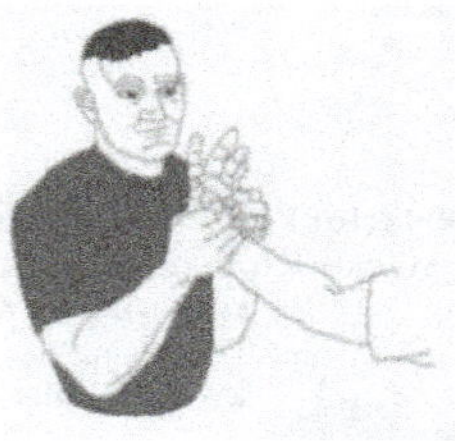

Step into the opponent and apply pressure downward on the back of his hand to bend the joint and rotate his wrist away from the body to twist the joint.

Enhanced Pain Compliance on Wristlock. Enhanced pain compliance techniques are applied in the third and fourth levels in the continuum of force. Additional pain be applied to a wristlock by-

Adding downward pressure to the elbow with the other hand or elbow by using the fingers to pull in on the opponent's radial nerve located on the inside of the forearm. When pressure is added to the opponent's radial nerve, his direction can be controlled.

Applying pressure against the opponent's finger joint to bend it in a direction it cannot bend (i.e., splitting the fingers).

Come-Along Holds

Marines use a come-along hold to control and move an opponent.

Escort Position. A common come-along hold is the escort position. To execute the escort position, Marines—

Face the opponent. Use the left foot to step forward at a 45-degree angle. Turn to face the right side of the opponent.

Use the right hand to firmly grasp the opponent's right wrist. With the left hand, firmly grasp the opponent's right triceps.

Position the opponent's controlled arm diagonally across the torso, keeping his wrist against the right hip. The Marine should be standing to the right of and behind the opponent.

Note: This technique works well when escorting an opponent on either the right or left side. Take caution when escorting an opponent by ensuring his controlled hand is not in a position to grab the holstered weapon. The preferred escort position is from the Marine's left side, so that the opponent is kept further away from the weapon.

Wristlock Come-Along. To execute the wristlock come-along, Marines-

Use the left hand to execute a basic wristlock. Incorporate the right hand in a two-handed wristlock for more control.

Maintain pressure on the opponent's wrist with the right hand, step forward, and pivot around to stand next to the opponent.

Release the left hand, quickly reach under the opponent's arm from behind, and grab his hand.

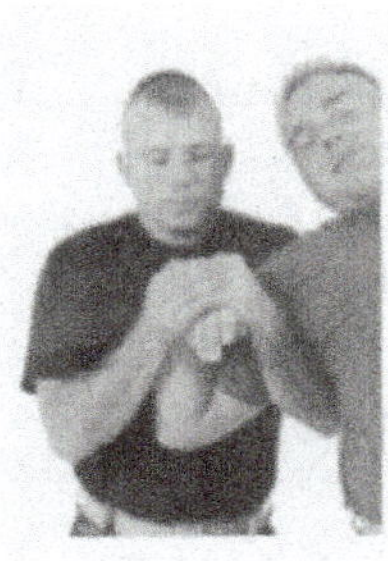

Use the left hand and apply downward pressure on the opponent's wrist.

Controlling Technique. The following controlling technique is used when an opponent grabs the Marine's wrist. To execute the technique, Marines—

Trap the opponent's hand with the palm of the other hand.

Rotate the opponent's trapped hand up and on his forearm while maintaining downward pressure on his trapped hand.

Apply downward pressure with both hands until the opponent is taken to the ground.

Armbars

An armbar is a joint manipulation in which pressure is applied on a locked elbow, just above the joint, in the direction the joint will not bend. An armbar has to be locked in quickly, but still requires a slow, steady pressure to gain compliance.

Basic Armbar. To execute a basic armbar, Marines-

Use the right hand to grab the opponent's right wrist.

Bring the left hand down on or above the opponent's elbow joint. To gain additional leverage, pivot to face the opponent.

Use the left hand to apply downward pressure on or above the opponent's elbow joint while pulling up on his wrist.

Armbar from a Wristlock. To execute an armbar from a wristlock, Marines-

Use the right hand to grab the opponent's right hand and execute a reverse wristlock.

Bring the left hand down on or above the opponent's elbow joint.

Use the left hand to apply downward pressure on or above the opponent's elbow joint while pulling up on his wrist.

Takedowns

A takedown is used to bring an opponent to the ground to further control him.

Takedown From a Wristlock Come-Along. To take the opponent to the ground from a wristlock come-along, Marines—

Use the right foot to push down on the opponent's calf or Achilles tendon.

Maintain control of the opponent's wrist and elbow and apply a slow, steady pressure to bring him to the ground.

Armbar to a Takedown. This technique is used to take a noncompliant opponent to the ground from an armbar. To execute the armbar takedown, Marines—

Use the right hand to execute a reverse wristlock.

Bring the left hand or forearm down on or above the opponent's elbow joint.

Pivot so the back is facing the opponent and, at the same time, lift the left elbow and slide the body so it is against the opponent, placing the armpit high above the opponent's elbow joint.

Lean back, placing the body weight on the opponent's arm until he complies or is taken to the ground.

Note: This technique may break the opponent's arm. Therefore, this technique should not be employed if the objective is a nonlethal takedown.

Wristlock Takedown. This technique is used to take a noncompliant opponent to the ground from a basic wristlock and to put him in a position where he can be handcuffed, if necessary. To execute the takedown, Marines-

Use the right hand to execute a basic wristlock.

Incorporate the left hand in a two-handed wristlock.

Apply downward pressure on the wristlock, pivot on the ball of the right foot, and quickly turn to the right to take the opponent to the ground.

Continue to apply pressure on the wristjoint as the opponent lands on his back with his arm straight in the air.

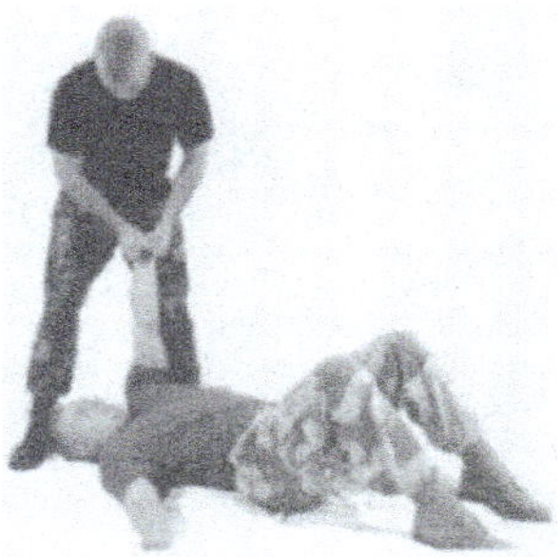

Slide the left foot under the opponent's back, underneath his armpit.

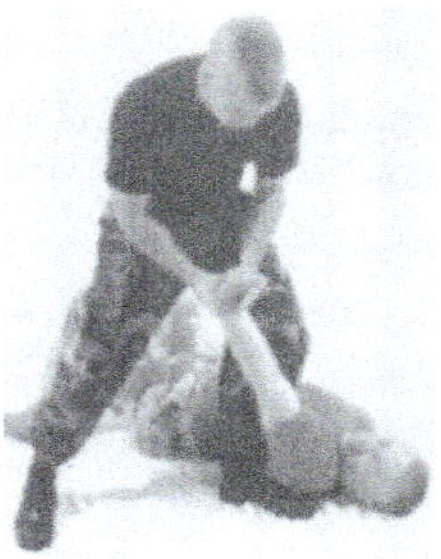

Apply pressure with the knee against the opponent's triceps while pulling back on his arm.

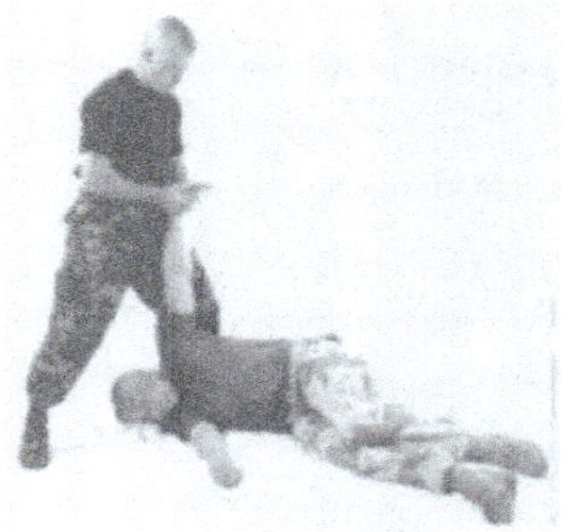

Continue applying pressure with the knee on the opponent's arm. Pivot and step around the opponent's arm to roll him on his stomach.

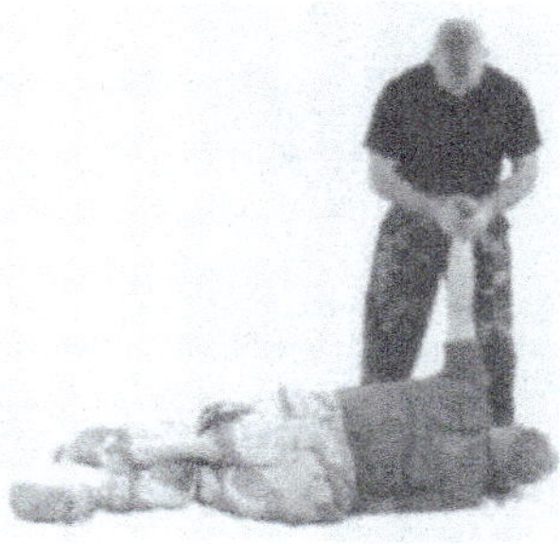

Kneel down with one knee on the opponent's back. The other knee is placed on the opponent's neck and shoulder on either side of his arm. Apply inward pressure with the knees to lock his arm in place.

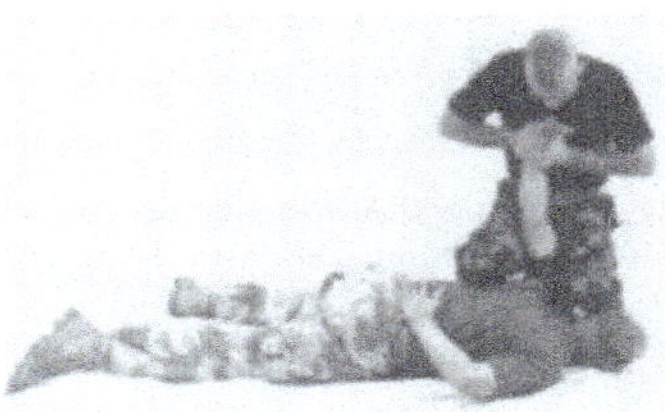

Tell the opponent to put his other hand in the middle of his back. Bring the opponent's controlled hand to the center of his back.

Escort Position Takedown. This technique can be used to control a noncompliant opponent from the escort position. To execute the escort position takedown, Marines—

Lock the opponent's arm straight across the body while rotating his wrist away from the body.

Use the left hand or forearm and apply downward pressure above the opponent's elbow where the triceps meet.

Step back with the right foot and, keeping the opponent's hand controlled against the hip, pivot to the right while continuing to apply downward pressure on his arm to bring him to the ground.

Note: This technique works well when escorting a subject on the Marine's right or left side. When taking down a subject from the right side, step back and pivot to the left.

2. Nonlethal Baton

A baton or nightstick can be an effective compliance tool when used correctly. Batons can be used defensively (blocking), offensively (striking), and as a restraining device when needed. In the fourth level of the continuum of force (assaultive [bodily harm]), defensive tactics include baton or nightstick blocks and blows. Blows to the head or other bony parts of the body are considered deadly force. When deadly force is not authorized, Marines must be able to employ blocks, strikes, and restraints with the baton with the minimum of force.

Grips

One-Handed Grip. To execute the one-handed grip, Marines—

Use the right hand to grasp the lower end of the baton, about 2 inches from the end.

Wrap the thumb and index finger around the baton so they are touching one another. The grip on the baton should be firm, but natural.

Two-Handed Grip. To execute the two-handed grip, Marines-

Use the right hand to grasp the lower end of the baton, about 2 inches from the end. Wrap the thumb and index finger around the baton so they touch one another.

Use the left hand to grasp the upper end of the baton, palm down, about 2 inches from the end. The hands should be approximately 10 to 12 inches apart.

Stance and Method of Carry

The basic warrior stance serves as the foundation for initiating nonlethal baton techniques. The method of cany provides effective defensive positions with a wide range of options to control a combative opponent.

One-Handed Carry. To execute the one-handed carry, Marines—

Grip the baton using the one-handed grip.

Elevate the baton, with the gripping hand at a level between the belt and shoulder.

Keep the left hand in the position of the basic warrior stance.

Two-Handed Carry. This carry is effective for blocks. To execute the two-handed carry, Marines-

Grip the baton using the two-handed grip.

Elevate the baton, with the left hand higher than the right hand.

Orient the weapon toward the subject.

Movement

In a nonlethal confrontation, movement may be made to create distance between Marines and the opponent or to close the gap to control him. When facing an opponent, Marines move in a 45-degree angle to either side of the opponent. Moving at a 45-degree angle is the best way to both avoid an opponent's strike and to put Marines in the best position to control the opponent.

Blocking Techniques

One-handed blocks are used when carrying the baton in the one-handed carry. The same one-handed blocks used in combative stick techniques (see page 3-6) apply in nonlethal baton engagements. The blocks include blocks for a vertical strike, a forward strike, and a reverse strike. Because the baton is often carried with two hands, there are also two-handed blocks that are effectively used from this carry. Two-handed blocks are discussed in the following subparagraphs.

High Block. Marines execute a high block to deter a downward vertical attack directed at the head and shoulders. To execute the high block, Marines—

Raise the baton up to a level even with or above the head. The baton should be in a horizontal position to block the blow. The fingers of the left hand should be open and behind the baton.

Place the baton perpendicular to the opponent's striking surface to absorb the impact of the blow.

Bend the elbows slightly to help absorb the impact of the blow. The arms should give with the strike of the blow.

Low Block. Marines execute a low block to deter an upward vertical attack directed at the abdomen, groin, or torso. The opponent's blow can be delivered by a foot, knee, or fist. To execute the low block, Marines—

Lower the baton to a level even with or below the groin. The baton should be in a horizontal position to block the blow. The fingers of the left hand should be open and behind the baton.

Place the baton perpendicular to the opponent's striking surface to absorb the impact of the blow.

Bend the elbows slightly to help absorb the impact of the blow. The arms should give with the strike of the blow.

Right Block. Marines execute a right block to deter an attack directed at the head, neck, flank, or hip. The opponent's blow can be delivered by a foot, knee, fist, or elbow. To execute the right block, Marines-

Thrust the baton in a vertical position to the right side.

Pivot to the right by stepping with the left foot and pivoting off the ball of the right foot. Rotate the hips and shoulders into the direction of the block. The fingers of the left hand should be open and behind the baton.

Place the baton perpendicular to the opponent's striking surface to absorb the impact of the blow.

Bend the elbows slightly to help absorb the impact of the blow. The arms should give with the strike of the blow.

Left Block. Marines execute a left block to deter an attack directed at the head, neck, flank, or hip. The opponent's blow can be delivered by a foot, knee, fist, or elbow. To execute the left block, Marines—

Thrust the baton in a vertical position to the left side.

Pivot to the left by stepping with the right foot and pivoting off the ball of the left foot. Rotate the hips and shoulders into the direction of the block. The fingers of the left hand should be open and behind the baton.

Place the baton perpendicular to the opponent's striking surface to absorb the impact of the blow.

Bend the elbows slightly to help absorb the impact of the blow. The arms should give with the strike of the blow.

Middle Block. Marines execute either a left or right block to deter an attack directed at the face, throat, chest, or abdomen. To execute the middle block, Marines—

Thrust the baton in a vertical position straight out in front of the body. The fingers of the left hand should be open and behind the baton.

Place the baton perpendicular to the opponent's striking surface to absorb the impact of the blow. The baton should be held with the left hand forward of the right.

Bend the elbows slightly to help absorb the impact of the blow. The arms should give with the strike of the blow.

Restraining Technique

The strong-side armlock is used to restrain an opponent who is not compliant. To execute the strong-side armlock, Marines—

Use the right hand to run the baton up under the opponent's left armpit, parallel to the ground.

Use the right foot to step forward at a 45-degree angle to the left side of the opponent. The baton is across his forearm.

Use the right hand to drive the baton forward and up so the action bends the opponent's arm behind his back. At the same time, continue moving around him to get behind him.

Place the baton on the opponent's forearm with the thumb and/or fingers and apply pressure to his forearm. At the same time, grasp the other end of the baton with the left hand.

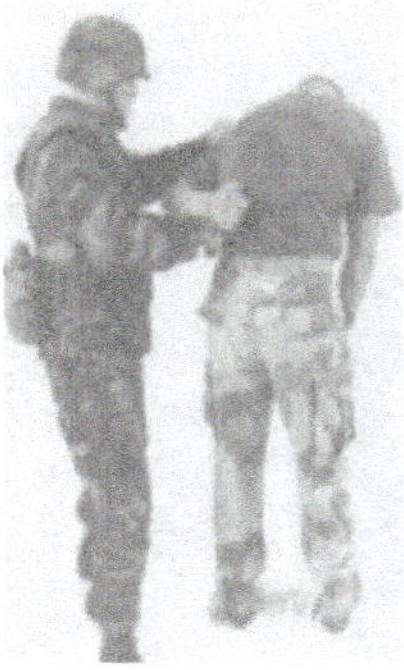

Pull up on the low end of the baton with the right hand. At the same time, push down on the top end of the baton with the left forearm, reaching around with the left hand to grasp the opponent's biceps or shoulder.

Continue exerting downward pressure with the left forearm while pulling back on the opponent's biceps with the left hand. This places the opponent in a position where he is controlled and can be moved.

Apply pressure with the foot against the bend in the opponent's leg above his calf. This lowers the opponent to the ground rather than throwing him to the ground and risking severe injury.

Striking Target Areas

Marines must avoid striking an opponent in the head, neck, or other bony parts with the baton because this is considered deadly force and can lead to serious bodily injury or death. Instead, the legs, arms, and buttocks are target areas that are considered nonlethal.

Legs. Primary targets are the thighs and lower legs. Avoid striking the knee and ankle joints because this can cause permanent damage.

Arms. Primary targets are the upper arms. Avoid striking the shoulder and elbow and wrist joints because this can cause permanent damage.

Buttocks. Primary targets are the buttocks. Avoid striking any other part of the torso, including the chest, rib cage, spine, tail bone, and groin because strikes to these areas can cause permanent damage or death.

One-Handed Striking Techniques

One-Handed Forward Strike. A forward strike follows either a horizontal line or a downward diagonal line using a forehand stroke. To execute the one-handed forward strike, Marines-

Stand facing the opponent with the baton carried in a one-handed cany.

Place the right hand palm up, swing the baton from right to left, and make contact with the opponent.

One-Handed Reverse Strike. A reverse strike follows either a horizontal line or a downward diagonal line using a backhand stroke. To execute the one-handed reverse strike, Marines—

Stand facing the opponent with the baton carried in a one-handed carry.

Bend the right arm and cross the arm to the left side of the body. The baton should be close to or over the left shoulder.

Place the right hand palm down, swing the baton from left to right, and make contact with the opponent.

Two-Handed Striking Techniques

Two-Handed Forward Strike. This strike is an effective follow-up to a middle block or left block. To execute the two-handed forward strike, Marines—

Pull back with the left hand while driving the right hand forward toward the opponent. The baton should be horizontal to the ground. Power is generated by stepping forward with the right foot and rotating the right hip and shoulder into the strike.

Contact the opponent with the end of the baton.

Two-Handed Reverse Strike. This strike is an effective follow-up to a middle block or left block. To execute the two-handed reverse strike, Marines—

Pull back with the right hand while driving the left hand forward toward the opponent. The baton should be horizontal to the ground. Power is generated by stepping forward slightly with the left foot and rotating the left hip and shoulder into the strike.

Contact the subject with the end of the baton.

Front Jab. This strike is effective for countering a frontal attack. It can also be executed as a quick poke to keep a subject away. To execute the front jab, Marines thrust both hands forward in a quick jab. The baton is held either horizontal to the ground or at a slight downward angle.

Rear Jab. This strike is effective for countering a bear hug from the rear. To execute the rear jab, Marines thrust both hands rearward in a quick jab. The baton is held either horizontal to the ground or at a slight downward angle.

APPENDIX A

PUGIL STICK TRAINING

A pugil stick is a training device used to simulate a rifle bayonet so that effective, but safe, training can be conducted to build proficiency of rifle bayonet techniques. Pugil stick training builds on the techniques used to throw punches. Pugil stick training is the only "full contact" training provided to Marines in the Close Combat Program. Pugil stick training teaches Marines to function when faced with stress and violence, and it prepares them to deliver a blow and take a blow. It also provides them with the physical and mental skills vital to success on the battlefield.

1. Pugil Stick Training

Design

A pugil stick consists of a stick wrapped in padding at both ends that can be gripped like a rifle. The pugil stick is approximately the same weight and length of an unloaded rifle with a bayonet attached.

Holding the Pugil Stick

The pugil stick is held in the same manner as the sen ice rifle. All movements come from the basic warrior stance. To hold the pugil stick correctly, Marines—

Use the right hand to grasp the lower end of the pugil stick overhanded.
Use the left hand to grasp the upper end of the pugil stick underhanded.
Use the right forearm to lock the lower end of the pugil stick against the hip.
Orient the blade end of the pugil stick toward the opponent.

Safety Equipment

The following safety equipment must be worn during any pugil stick bout.

Groin Protection. Groin protection protects the groin from an accidental blow. It should be pulled high around the waist to protect the groin area, with the concave portion against the body.

Flak Jacket. The flak jacket provides protection to the body. It is worn completely fastened.

Neck Roll. The neck roll prevents whiplash if Marines receive a blow to the head. The neck roll further supports the head and protects the neck from blows. The neck roll is worn above the flak jacket and below the helmet. The tied end of the neck roll faces the front.

Helmet. A regulation football helmet protects the face and head. The helmet must fit snugly and the chin straps must be adjusted and snapped.

Mouthpiece. The mouthpiece is worn on the upper teeth to protect the teeth.

Gloves. Marines wear gloves to protect the hands if pugil sticks do not have gloves built-in.

Pugil Stick Screening

Prior to pugil stick training, instructors must ask participating Marines the following questions. If a Marine answers "yes" to any question, he must see the corpsman or a competent medical authority who evaluates him and determines whether or not the he can participate in the training.

- Are you on light duty?
- Arc you restricted to running shoes by a corpsman or a doctor?
- Have you fought in a pugil stick bout within the last 7 days?
- Have you received a blow to the head within the last 7 days?
- Have you had a concussion within the last 6 months?
- Have you had dental surgery within the last 24 hours?
- Do you have stitches or staples on your body?
- Have you had a shoulder or head injury within the last 5 years?
- Are you taking a prescription drug?
- Do you have an ear infection or current sinus infection?
- Have you had a broken bone within the last 6 months?

Safety Personnel

The following safety personnel are required to conduct pugil stick training:

- One close combat instructor must officiate the bout.
- One close combat instructor trainer, commissioned officer, or staff noncommissioned officer must be in the training area to serve as range safety officer.
- One corpsman must be in the training area.

Training Area

To prevent injury, Marines train on areas with soft footing (i.e., sand or grass). Training mats are not recommended because feet can stick to the mats, prohibiting movement or causing joint injuries by twisting a knee. Bouts should not take place on a hard surface area; e.g., a flight deck or parking lot. A boxing ring may be used to conduct pugil stick bouts; ring dimensions can vary as long as there is ample room to execute the techniques in the training area.

Second Impact Syndrome

Second Impact Syndrome occurs when a second blow to the head produces a second concussion that occurs within 1 week following a previous concussion (before recovery from the first concussion). Second Impact Syndrome causes rapid brain swelling and can cause death. Therefore, there must be 7 days between pugil stick bouts to reduce the risk of severe injury resulting from Second Impact Syndrome. The 7 day separation between pugil stick bouts significantly reduces the possibility of injury, particularly in someone who may have suffered a brain injury or concussion but shows no symptoms.

Any Marine who experiences headaches or the following symptoms after training must be examined by appropriate medical personnel:

- Blurred vision.
- Ringing in the ears.
- Dilation of the pupils.
- Slurred speech.
- Bleeding from ears or mouth.
- Swelling in head or neck area.
- Any unnatural discoloration in head or neck area.

The Marine should not be allowed to participate in pugil stick training or any other activity where a heavy blow might be sustained for a minimum of 7 days after the headache or other symptoms have subsided.

Safety Measures

The following safety measures must be followed:

- Gear must be worn properly throughout training.
- Contact lenses or glasses will not be worn.
- False teeth will be removed from the mouth.
- Nothing will be worn around the neck except the neck roll.

Competition among groups of Marines is authorized as long as it does not overshadow training objectives or compromise safety procedures.

Safety and proper techniques are paramount.

Safety is more important than competition.

2. General Rules and Regulations Governing Pugil Stick Bouts

Instructors and Support Personnel Requirements

At a minimum, one close combat instructor observes a pugil stick bout. For safety purposes, it is better to have two instructors judging a bout because each instructor can fully observe each of the fighters. The best position for observation is to the right of a fighter. This allows the instructor to see the fighter's facial expression and body movement. The instructor's position must not interfere with the fight.

In addition, one commissioned or staff noncommissioned officer (to serve as range safety officer) and one corpsman will be in the training area.

Prior to the Bout

Prior to the bout, Marines—

Are paired according to height, weight, and gender.

Wear the proper safety equipment.

Once Marines have properly donned the safety equipment, they wait for instructions from the close combat instructor. On the command of the close combat instructor, a pair of Marines will enter the ring. Once in the ring, the close combat instructor inspects each Marine for the proper safety equipment.

Note: There should he enough gear so that when two Marines are training, two other Marines can be donning safety equipment for the next bout.

During the Bout

The bout begins when the close combat instructor blows the whistle. All strikes are directed above the waist. Upon hearing a whistle blast, all fighting immediately ceases.

Stopping a Bout

There arc two reasons for stopping a bout: delivery of a scoring blow or an unsafe condition. The close combat instructor trainer, close combat instructor, or the range safety officer may stop the bout at any time an unsafe condition is observed.

Scoring Blow. A scoring blow is an offensive technique delivered to a vulnerable area of an opponent with sufficient force and precision to be considered as a disabling or killing blow. Scoring blows are not judged solely on the degree of force with which a blow is delivered, but on the accuracy and techniques employed. A scoring blow is defined as—

A straight thrust with the blade end of the weapon (red end of the pugil stick) to the opponent's face mask or throat.

A slash to the side of the opponent's helmet (below the ear) or neck with the red end of the pugil stick.

A heavy blow to the opponent's head with an authorized technique (i.e., buttstrokc, smash) using the butt (black end) of the pugil stick.

When a scoring blow is delivered, the close combat instructor blows the whistle to stop the bout.

Unsafe Condition. The bout will be stopped as soon as an unsafe condition exists. An unsafe condition exists when a Marine is unable to defend himself, loses his balance and falls down on one or two knees or falls down completely, shows instability (e.g., buckling at the knees), loses muscular tension in his neck and his head snaps back or to one side, or appears disoriented. An unsafe condition also exists if a Marine lets go of one end of the pugil stick, the equipment (e.g., helmet, neck roll) falls off, or a Marine fails to use the proper techniques.

If any of these conditions occur—

The close combat instructor trainer, close combat instructor, range safety officer, or anyone supervising training stops the bout and separates the two Marines.
The corpsman evaluates the possibly injured Marine to determine if the fight can continue. The corpsman—

- Ensures that he is alert and responsive, both verbally and physically.
- Talks to him to see if he is coherent.
- Makes certain he comprehends and replies to verbal questions.
- Makes certain his speech is not slurred.
- Checks his physical signs.
- Ensures his eyes are focused and not dazed or glazed.
- Ensures his legs are not wobbly or shaky.

The pugil stick is not used as a baseball bat. The use of unauthorized techniques will result in expulsion from the bout. Only techniques taught in bayonet training are authorized. These techniques include-

Straight thrust.
Buttstroke (horizontal and vertical).
Smash.
Slash.
Parry.
Blocks (low, high, left, right).

After the Bout

A third whistle blast is used prior to identifying the Marine who delivered the killing blow.

A fourth whistle blast is used prior to demonstrating the killing technique used by the Marine who won the bout.

3. Directions for Making Pugil Sticks

Close combat instructor trainers are taught to make pugil sticks to standard specifications at school. It is recommended that all pugil sticks be made under the supervision of a close combat instructor trainer. Until the Marine Corps acquires a standardized pugil stick that units can order through the supply system, instructors use the following information and instructions to make pugil sticks locally.

Supplies

The following supplies are needed to make one pugil stick:

One 45" by 2" circular, oak, wood dowel.
Four 9" wide by 27" long foam pad pieces to make two inner handguards of double thickness.
Four 11" wide by 19" long foam pad pieces to make two outer handguards of double thickness.
Canvas material to cover inner and outer handguards.
One foam pad cut 1/4" thick by 7" wide by 13" long to make the centerguard.
One 8" wide by 14"long piece of canvas material to make a pugil stick sock.
Canvas material to make two pugil stick end caps. Ideally, two different color fabrics should be used to simulate the bayonet and buttstock ends of a rifle. Common colors for end caps are black and red. Each end cap is a different color to help the officiating certified close combat instructor determine the scoring blow during the pugil stick bout.
Two 1/4" thick by 11" wide by 54" long foam pad pieces to make two ends.
Two foam pads cut 1/4" thick by 3" wide by 14 1/2" long to make two end plug inserts.
One can of 3M Photo MountQ spray adhesive (NSN 8040-01-120-4009)to adhere foam padding to the stick. (Spray adhesive is available from self-service.)
One roll of duct tape or riggers tape (NSN 7510-00-266-5016) to protect foam pad and

canvas from wear and to reinforce gluing action of spray adhesive. (Duct tape or riggers tape is available from self-service.)
One roll para-cord or 5-50 cord to tie both the inner and outer handguards together.
One roll of thread strong enough to sew canvas material together.
Sixteen grommets, one for each of the four corners of the inner and outer handguards.
One foam sleeping mat (NSN 8465-01-109-3369). Use unserviceable foam sleeping mats obtained from Defense Rcutilization and Marketing Office.

Tools

The follow ing tools are needed to make a pugil stick:

Tape measure, ruler, or yardstick.
Scissors.
Grommet machine.
Knife.
Sewing machine.

Directions

To make a pugil stick, Marines perform the following steps.

Preparing Materials

Step 1. Cut foam sleeping mat into the dimensions listed above. Cut mats to double thickness.

Step 2. Use spray adhesive to adhere the two 9" by 27" pieces of foam together. Use the spray adhesive to adhere the two 11" by 19" pieces of foam together. Let adhesive dry before using the double-thick foam for handguards.

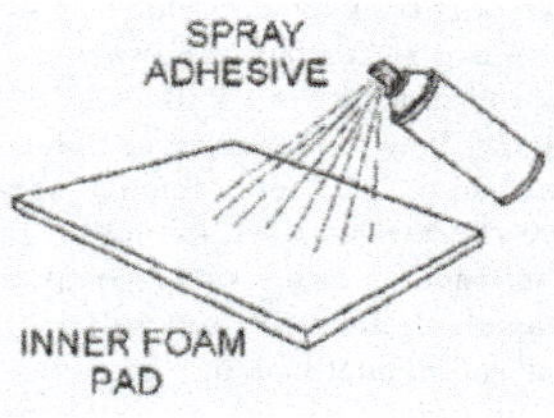

Step 3. Cut canvas material to make two coverings for the inner handguards. Cut material to make two coverings for the outer handguards. Two inner handguards and two outer handguards are required for each pugil stick.

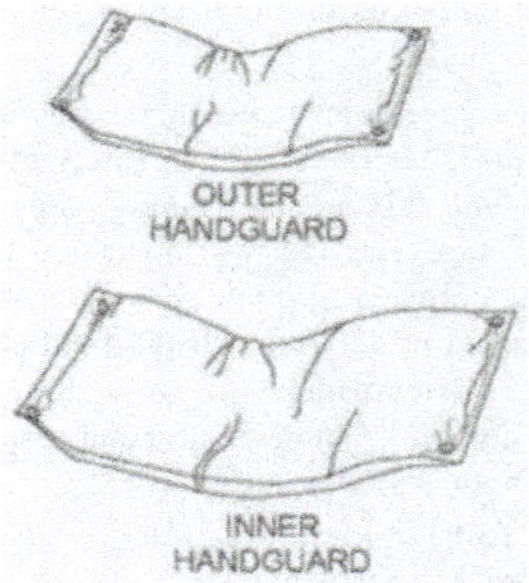

Step 4. Use a sewing machine to make a sleeve out of the 9" by 27" canvas. Sew down three sides. Insert foam pad inside the canvas sleeve and then sew down the fourth side. Use the grommet machine to create holes reinforced with metal grommets in each of the four corners. Once finished, the inner handguards arc complete.

Note: It is recommended that the Fabric Repair Shop be used to sew canvas sleeves and apply grommets.

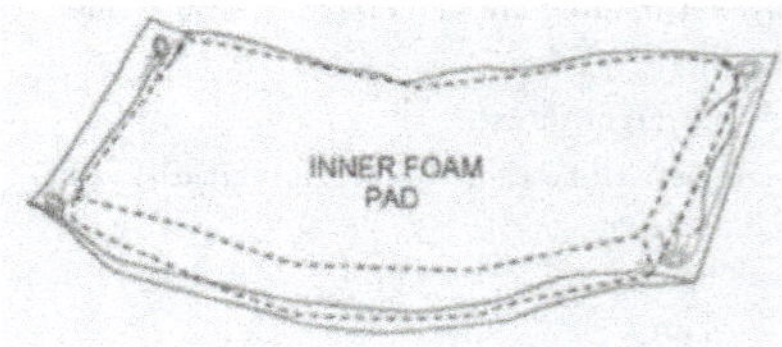

Step 5. Repeat step 4 to make the other inner handguard and both outer handguards.

Constructing Ends, End Plugs, and End Caps

Step 1. Apply spray adhesive on approximately 8 inches of one end of the dowel. This adhesive serves to glue the foam pad to the wooden dowel.

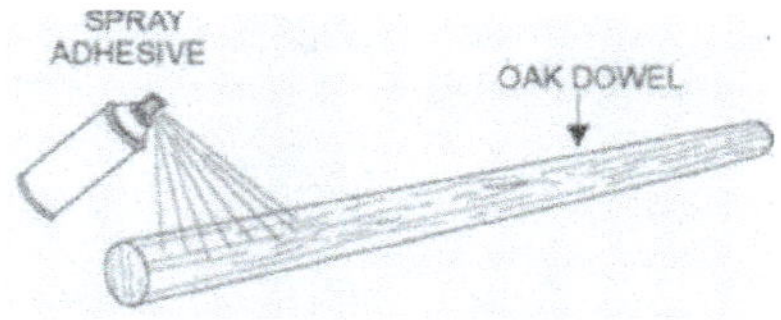

Step 2. Apply spray adhesive on one side of the 11" by 54" foam pad. This causes the foam pad to stick to itself when it is rolled around the end of the wooden dowel. Ensure 3 inches of the rolled foam pad extends past the wooden dowel for stability.

Step 3. Apply spray adhesive on one side of 3" by 14 1/2" foam pad. Roll pad into a tight cylinder.

Step 4. Apply spray adhesive on the exposed end of the dowel and fill the hole in the end with the end plug. The end plug prevents the wooden dowel from protruding beyond the edge of the rolled foam padding. This is an important safety precaution to prevent injury.

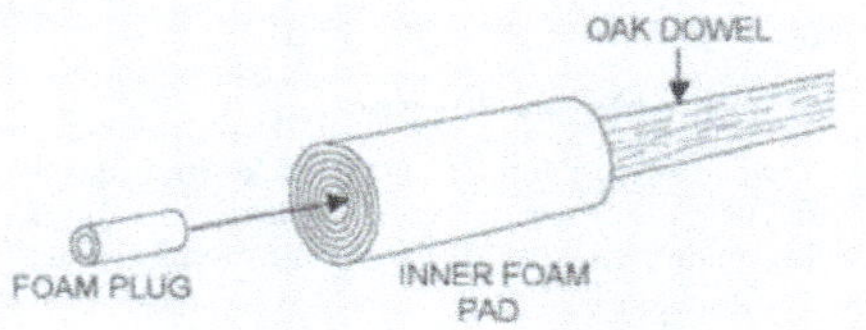

Step 5. Use duct tape or riggers tape to secure and reinforce the foam pad once the end plug is inserted in the end.

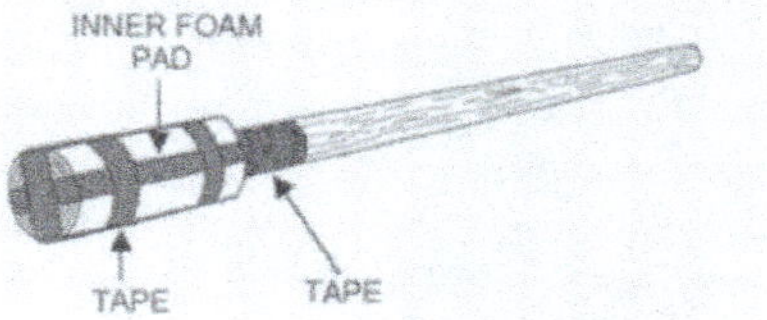

Step 6. Measure the dimensions of the ends (typically, the length of the cylinder should be 12 inches long to allow for taping). Cut and sew canvas material in the shape of a cylinder with one open end. Once complete, this becomes the end cap. The end cap is a colored sleeve that covers the rolled and taped foam pad at each end of the pugil stick.

Note: It is recommended that the Fabric Repair Shop be used to make the end cap.

Step 7. Slide one of the end caps over the rolled foam pad and tape it to the dowel.

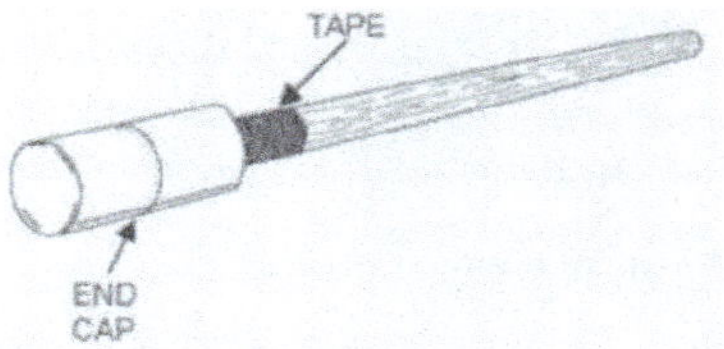

Building the Centerguard

Step 1. Spray one side of the 7" by 13" foam pad with spray adhesive and wrap it tightly around the center of the dowel.

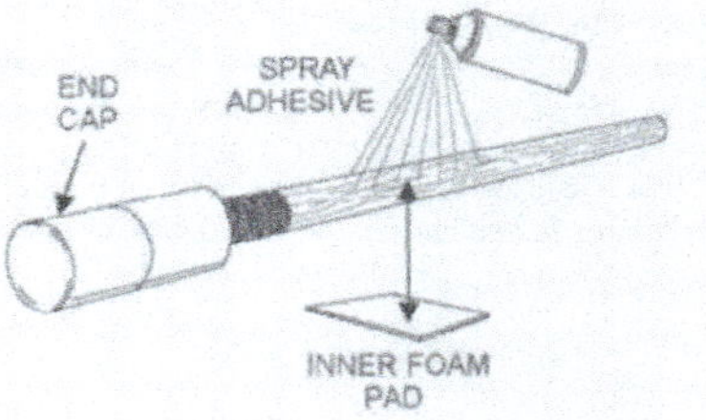

Step 2. Wrap duct tape or riggers tape around the foam pad to secure.

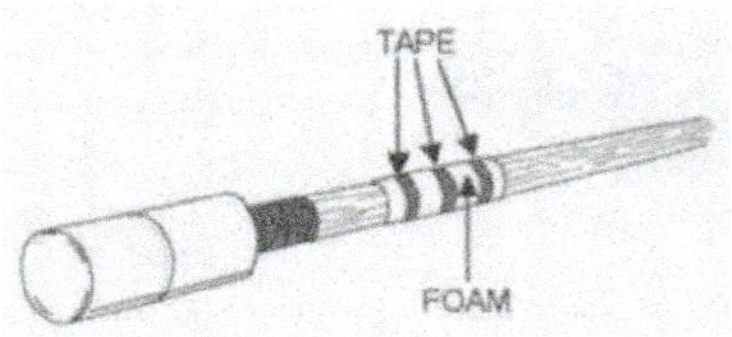

Step 3. Use a sewing machine to make a tubular pugil stick sock out of the 8" by 14" piece of canvas. Slide the pugil stick sock over the rolled foam pad and secure it with the duct tape or riggers tape. Duct tape or riggers tape protects the edges of the centerguard and reinforces the gluing effects of the spray adhesive.

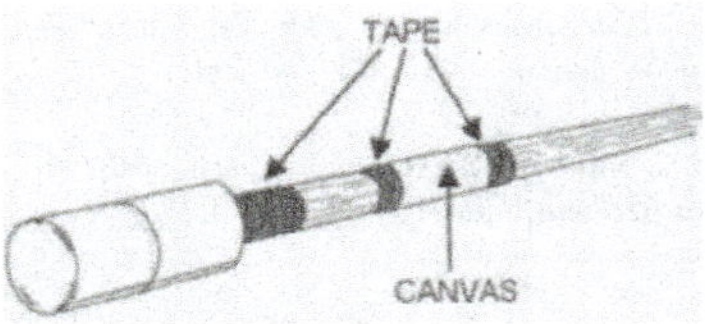

Note: It is recommended that the Fabric Repair Shop be used to make the pugil stick sock.

Finishing the Opposite End Cap

After one end cap is complete and the centerguard is finished, then repeat steps 1 through 7 in Constructing Ends, End Plugs, and End Caps for the other end of the pugil stick. Ensure that the opposite color end cap is used.

Fastening Handguards

Step 1. Place the inner handguards (longer piece) around the dowel. The inner handguard is placed on first because it protects the wrist. One inner handguard is needed for the left hand and one for the right hand.

Step 2. Fold the inner handguard around the pugil stick and run the para-cord or 5-50 cord through the metal grommets.

Step 3. Tighten down the corner ends of the inner handguards by pulling on the para-cord or 5-50 cord.

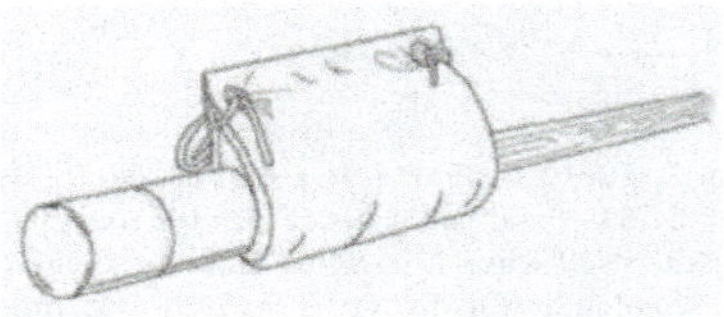

Step 4. Tie off the cord with a square knot above the metal grommets.

Step 5. Place the outer handguards (shorter piece) over the inner handguards (longer piece). Repeat steps 1-4 for the other handguard. The outer handguards provide additional protection to the knuckles and fingers and provides additional stability for the inner handguards.

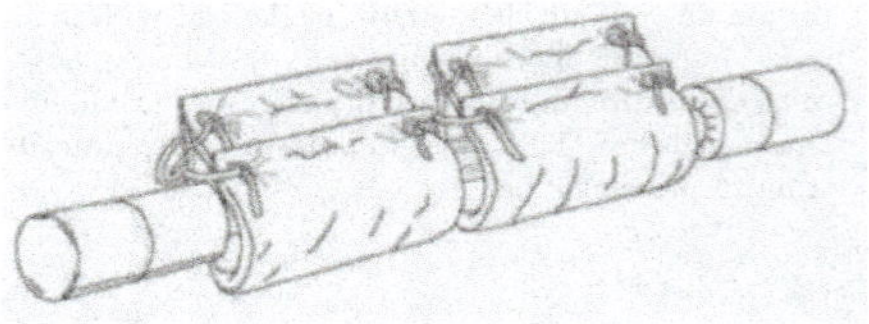

Serviceability Inspection

Prior to any training, all pugil sticks are inspected to ensure they are fit for use. If any of the following conditions exist, the pugil stick is unsafe and will not be used:

Stick protrudes through the end caps.
Frayed or worn edges exist on canvas surfaces.
The stick can be felt through the pads or the padding is too soft.
End caps or handguards are not securely attached.

(reverse blank)

APPENDIX B

SAFETY PRECAUTIONS DURING TRAINING

1. General Safety Precautions

When training close combat techniques, certain safety precautions must be adhered to to prevent injuries.

Most training should be conducted on a training area with soft footing such as a sandy or grassy area. If training mats are available, they should be used. A hard surface area is not appropriate for close combat training.

All techniques should be executed slowly at first. Marines can increase the speed of execution as they become more proficient. Marine-on-Marine training that requires contact (chokes, throws, ground fighting, and unarmed restraints and manipulation) should not be executed at full force or full speed.

If a technique is applied to the point that a Marine is uncomfortable, the Marine must "tap out." This indicates immediate release of the pressure being applied or to immediately stop the technique. The Marine "taps out" by firmly tapping his hand several times on any part of the opponent's body that will get his attention or by saying *stop*.

Second Impact Syndrome occurs when a second concussion develops within hours, days, or weeks following a prior concussion (and before recovery from the first concussion). Second Impact Syndrome causes rapid brain swelling and can cause death. Marines who experience headaches or other symptoms following training must be examined by appropriate medical personnel. These symptoms can include, but are not limited to, blurred vision, ringing in the ears, dialation of the pupils, bleeding from the ears or mouth, slurred speech, swelling in head or neck area, or any unnatural discoloration of head or neck. They should not be allowed to participate in pugil stick training or any other activity where a heavy blow might be sustained for a minimum of 7 days after the symptoms have subsided.

2. Safety Precautions for Individual Techniques

Falls

When training Marines to fall, they should progress from the ground, to a kneeling or squatting position, and then to a standing position. This ensures they are comfortable and understand the technique before progressing to executing falls from a higher profile. This instructional technique vastly reduces the risk of injury.

Strikes and Punches

When training Marines to strike and punch, they begin by executing the techniques "in the air." As they become more proficient, they execute strikes on equipment (when available) such as an air shield or a heavy bag. At no time should they be permitted to execute strikes on another student.

Chokes and Ground Fighting

Chokes

When training Marines to execute chokes, they will not apply pressure to the opponent's throat during training because the trachea and windpipe can be crushed. During training, Marines should practice the proper procedures for blood chokes. They should never execute air chokes. At no time during training should a choke be applied at full force or full speed. No choke should be held for longer than 5 seconds.

Unarmed Restraints and Manipulation

When training Marines to execute unarmed restraints and manipulation techniques, they utilize slow and steady pressure. Never apply these techniques at full force or at full speed.

Pugil Stick Training

A blow to the head during training will bruise the brain. A second blow to the head can cause death. This is known as the Second Impact Syndrome. There must be 7 days between pugil stick training to prevent injury or death.

UNITED STATES MARINE CORPS
Marine Corps University

SCRS 1201
26 Mar 2001

USMC Land Navigation

Sergeants Course

LEARNING OBJECTIVES.

1. Terminal Learning Objective. With the aid of references and given an area of operations during day/night, military map, protractor, compass, and grid coordinates, navigate using basic techniques, per FM21-26. (SCRS 12.01)

2. Enabling Learning Objectives.

a. Without the aid of references and given a grid coordinate, state the distance in meters that the grid coordinate will locate a point within, per FM21-26 (SCRS 1201 01a)

b. Without the aid of references and given a military map, protractor, and ten (10) eight (8) digit grid coordinates, plot the points on the map, per FM21-26 (SCRS 1201.01b)

c. Without the aid of references and given a military map, protractor, and ten (10) eight (8) digit grid coordinates, measure the distance from one point to another, per FM21-26 (SCRS 1201.01c)

d. Without the aid of references and given a military map, protractor, and ten (10) eight (8) digit grid coordinates, determine the grid azimuth between two points, per FM21-26 (SCRS 1201.01d)

e. Without the aid of references state the procedures used to navigate using a lensatic compass, per FM21-26 (SCRS 1201.01e)

f. Without the aid of references and given a military map, protractor, and back azimuths from two (2) known points, plot a resection, in order to identify your position on the map, per FM21-26 (SCRS 1201.01f)

g. Without the aid of references, state the procedures used to by-pass obstacles, per FM21-26 (SCRS 1201.01g)

h. With the aid of references and given a checklist, area of operations, lensatic compass, military map of the area, eight (8) digit grid coordinates, starting point, finishing point, and as a member of a two (2) man team, navigate during daylight hours in order to locate five (5) checkpoints on a map of the area in a time limit of four (4) hours, per FM21-26. (SCRS 1201.01h)

i. With the aid of references and given a checklist, area of operations, lensatic compass, military map of the area, eight (8) digit grid coordinates, starting point, finishing point, and as a member of a two (2) man team, navigate during darkness hours in order to locate three (3) checkpoints on a map of the area in a time limit of four (4) hours, per FM21-26 (SCRS 1201.01i)

OUTLINE:

1. COMPONENTS OF A MAP. A map can be compared to any other piece of equipment. Before you use it, you must read the instructions. Map instructions are placed around the outer edge of the map and are called "marginal information." A topographical map is defined as a type of map which portrays terrain features, as well as the horizontal positions of the features represented. The vertical positions, or relief, are normally represented by contour lines. All maps are not the same, so every time you use a different map, examine the marginal information carefully.

a. Sheet Name. A map is named after the most prominent cultural or geographical feature. Whenever possible, the name of the largest city on the map is used. The sheet name is found in two places: the center of the upper margin and either the right or left side of the lower margin.

b. Sheet Number. The sheet number is used as a reference number for that map sheet. It is found in two places: the upper right margin and the lower left margin.

c. Series Name. The map series name is found in the upper left margin. It usually includes a group of similar maps at the same scale and/or the same sheet lines designed to cover a particular geographical area. The name given a series is that of the most prominent areas.

d. Scale. The scale note is a representative fraction that gives you the ratio of a distance on the map to the corresponding distance on the earth's surface. For example, the scale note 1:50,000 on many maps indicates that one inch on the map equals 50,000 inches on the ground. Maps with different scales will display different degrees of topographical detail. For example, a map with a scale of 1:25,000 will give more detail than a 1:50,000 map because one inch on the map represents only 25,000 inches on the ground, rather than the 50,000 inches of the 1:50,000 map. The scale is found both in the upper left margin after the series name and in the center of the lower margin.

e. Series Number. The series number is a sequence reference expressed either as a four-digit number (1125) or as a letter, followed by a three or four-digit number (M556; N3341). It is found in both the upper right margin and the lower left margin.

f. Edition Number. The edition number represents the age of the map in relation to other editions of the same map and the agency responsible for its production. For example, EDITION 5-DMATC indicates the fifth edition prepared by the Defense Mapping Agency Topographic Center. EDITION 6-DMA indicates the sixth edition prepared by the Defense Mapping Agency. Edition numbers run consecutively. A map bearing a higher edition number is assumed to contain more recent information than the same map bearing a lower edition number. It is found in the upper right margin and in the lower left margin.

g. Index To Boundaries. The index to boundaries diagram, which is a miniature of the map, shows the boundaries that occur within the map area, such as county lines and state boundaries. The index of boundaries diagram appears in the lower right margin of all map sheets.

h. Adjoining Sheets Diagram. Maps contain a diagram that illustrates the adjoining map sheets. The adjoining sheets diagram usually contains nine rectangles, but the number may vary depending on the locations of the adjoining sheets. All represented sheets are identified by their sheet numbers. The diagram is found in the lower right margin of all map sheets.

i. Elevation Guide. The elevation guide is a miniature characterization of the terrain shown. The terrain is shown by bands of elevation, spot elevations, and major drainage features. The elevation guide helps you rapidly identify major land forms. It is normally found in the lower right margin.

j. Declination Diagram. This indicates the angular relationships of true north, grid north, and magnetic north. Recent edition maps have a note indicating how to convert azimuths from grid to magnetic and from magnetic to grid next to the declination diagram. The declination diagram is located in the lower margin.

k. Bar Scales. Bar scales are used to convert map distance to ground distance. Maps may have three or more bar scales, each in a different unit of measure. Exercise care when using the scales, especially in the selection of the unit of measure. The bar scales are located in the center of the lower margin.

l. Contour Interval Note. The contour interval note states the vertical distance between adjacent contour lines on the map. When supplementary contours are used, the interval is indicated. In recent edition maps, the contour interval is given in meters instead of feet. This note is found in the center of the lower margin normally below the bar scales.

m. Legend. The legend illustrates and identifies the topographic symbols used to depict some of the more prominent features on the map, such as railroad tracks, buildings, and swamps. The symbols are not always the same on every map. Always refer to the legend to avoid error when reading a map. The legend is located in the lower left margin.

2. LOCATE GRID COORDINATES ON A MAP.

a. Grid lines. Grid lines are a series of straight lines intersected at right angles and forming a series of squares. It furnishes the map reader with a system of squares similar to the block system of most city streets. Two digits are printed in large type at each end of the grid lines, and these same two digits appear at intervals along the grid lines on the face of the map. They are called principal digits. They are of major importance to the map reader because they are the numbers he will use most often for referencing points.

b. Vertical Grid Lines. These run from the bottom of the map sheet to the top of the sheet (grid south---grid north). They function as the left and right (eastern and western) boundaries of a grid square and are labeled in the margin.

c. Horizontal Grid Lines. These run from the left side of the map sheet to the right side of the map sheet (grid west---grid east). They function as the bottom and top boundaries (southern and northern) of a grid square and are labeled in the margin.

d. Grid Squares. These intersect at right angles of the horizontal and vertical grid lines. The most common military map contains grid squares that measure 1000 meters by 1000 meters (not 1000 square meters as many people think). Any point located within the grid square is considered to be part of the grid square.

e. Basic Map Reading Rule. The designation of a point is based on the principle: Read right then up. Always read right on the vertical grid lines then up on the horizontal grid lines.

f. Grid Square Identification. It is important that all of you understand how to apply the map reading rule to identify a grid square and locate a point within a grid square.

(1) The coordinates of a grid square are found by combining the values of the vertical and horizontal grid lines that form the lower left-handed corner of that grid square.

(2) First read right to the vertical grid line that forms the left (western) boundary of the grid square and record the principal digits.

(3) Next read up on the horizontal grid line that forms the left (western) boundary of the grid square and record the principal digits.

(4) The combination of the principal digits that label the vertical grid line and horizontal grid line are the identification of the grid square or its coordinates.

(5) A four digit grid coordinate locates a point to within 1000 square meters, on the map, which is called a grid square.

(6) A six digit grid coordinate will locate a point on a map within 100 meters.

(7) An eight digit grid coordinate will locate a point on a map within 10 meters

3. PLOTTING POINTS ON A MAP.

a. 6-Digit Grid Coordinates. Imagine dividing a grid square into 100 smaller squares. The coordinates of a point in such a grid square will have six digits (numbers). Each of the grid squares in figure 1201-1 is 1,000 meters long and 1,000 meters high. One of the grid squares is divided into 100 smaller squares--each 100 meters long and 100 meters high. Note that the lines within the grid square also read RIGHT, then UP. In a 6-digit grid coordinate, such as 284936, the first 3 numbers are the "read right" part of "read right, then up;" whereas, the last 3 numbers are the "then up" part. To locate point C in grid square 3050 (see figure 1201-1), you should use the following procedure:

(1) Read right to vertical line 30. Note that point C is 3/10 of the distance toward the vertical line 31. Write the vertical coordinate for point C as 303.

(2) Read up vertical line 30 to horizontal line 50. Note that point C is 7/10 of the distance toward the horizontal line 51. Write the horizontal coordinate of point C as 507.

(3) To write the 6-digit coordinate that locates point C, simply combine the two readings--303507. The vertical reading is ALWAYS placed before the horizontal reading.

NOTE: Do not actually draw the lines within the grid square (see figure 1201-1), since such lines would obstruct other information. You can determine a 6-digit grid coordinate by approximation, or by using your coordinate scale on a protractor. Use of the coordinate scale is explained in the next paragraph.

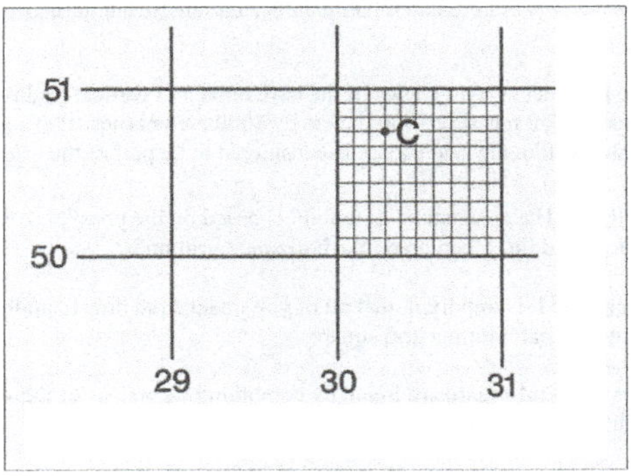

Figure 1201-1.

b. Protractor. If you require a more accurate 6-digit grid coordinate than can be obtained by estimation, you should use a coordinate scale. There are three coordinate scales located on your protractor: 1:100,000, 1:50,000, and 1:25,000. Use the one that corresponds with the scale of the map you are using. In most cases this will be 1:50,000. To locate point C in grid square 3050 using the coordinate scale on your protractor, use the following procedures:

(1) Place the proper coordinate scale of your protractor on the map so that the zero-zero point is to the bottom right of the scale (see figure 1201-2).

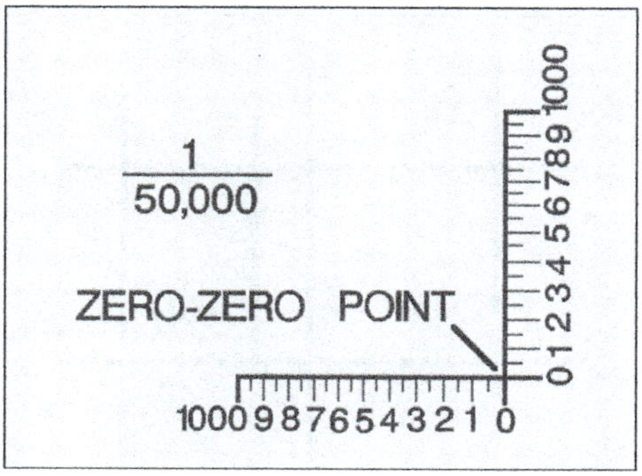

Figure 1201-2

(2) Place the zero-zero point at the lower left hand corner of grid square 3050 (see figure 1201-3).

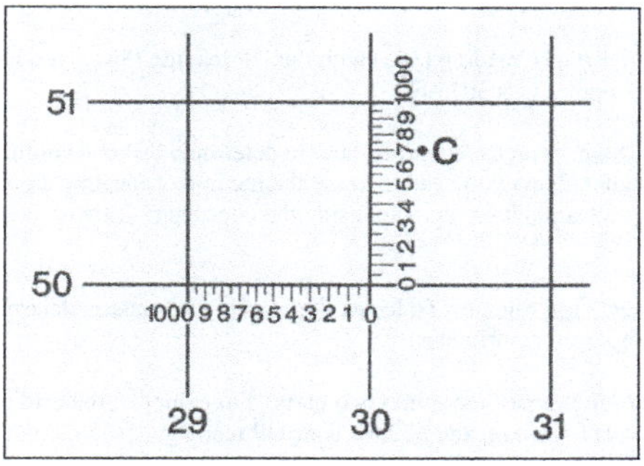

Figure 1201-3

(3) Keeping the horizontal line of the scale directly on top of the east-west grid line (in this case, grid line 50), slide it to the right until the vertical line of the scale touches the point (point C) for which the coordinates are desired (see figure 1201-4)

(4) Examine the two sides of the coordinate scale to ensure that the horizontal line of the scale is aligned with the east-west grid line, and the vertical line of the scale is parallel with the north-south grid line.

(5) Determine your RIGHT reading by first reading the value of the grid line to the left of point C (30). Add to this value the number that tells how far into the grid square point C is. In this case, it is 300 meters (see figure 1204-4). You now have the complete right reading of 303.

(6) Next , determine your UP reading by first reading the value of the horizontal grid line below point C (50). Add to this value the number which tells how far point C is up in the grid square. In this case it is 700 meters. You now have the complete UP reading of 507 (see figure 1201-4). When determining both your right and up reading, round your value to the closest number on your coordinate scale.

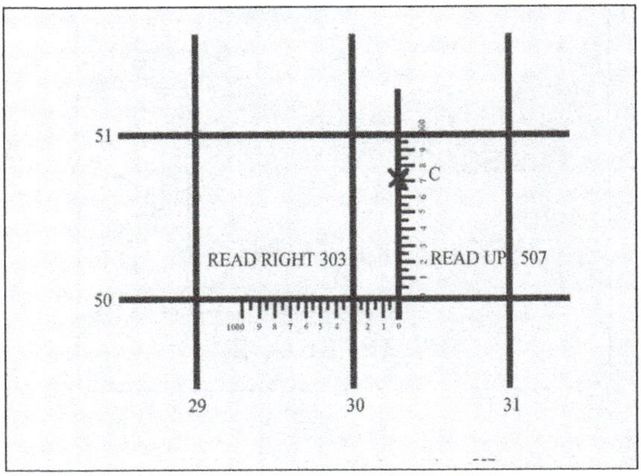

Figure 1201-4.

(7) By combining the RIGHT reading (303) with the UP reading (507), you have accurately determined that the 6-digit grid coordinate of point C is 303507.

c. Locating Points On A Map. You have learned how to determine the grid coordinates of a given point on a map to within 100 meters (6-digits). Next you need to know the method of locating a point on your map when given a 6-digit grid coordinate. To locate a point on the map, using the coordinate scale on your protractor, use the following procedures:

(1) Locate The Correct Grid Square. To locate the correct grid square, determine the 4-digit grid coordinates from the given 6-digit grid coordinates.

(a) Split your 6-digit coordinates into two parts. For example, the grid coordinates 025672 would be split to read 025 for your RIGHT reading, and 672 for your UP reading.

(b) Determine the vertical (north-south) grid line. It is the first two numbers of your RIGHT reading.

02

(c) Determine the horizontal (east-west) grid line. It is the first two numbers of your **UP** reading.

67

(d) Determine your 4-digit grid coordinates. Combine the **RIGHT** with the **UP** reading.

02 + 67 = 0267

(2) Plot The 6-Digit Coordinates. The procedures for plotting your grid coordinates are listed as follows:

(a) Place the proper coordinate scale of your protractor with the zero-zero point at the lower left hand corner of the grid square 0267, keeping the horizontal line of the scale directly on top of the east-west grid line.

(b) Examine the two sides of the coordinate scale to ensure that the horizontal line of the scale is aligned with the east-west grid line, and the vertical line of the scale is parallel with the north-south grid line (see figure 1205-5).

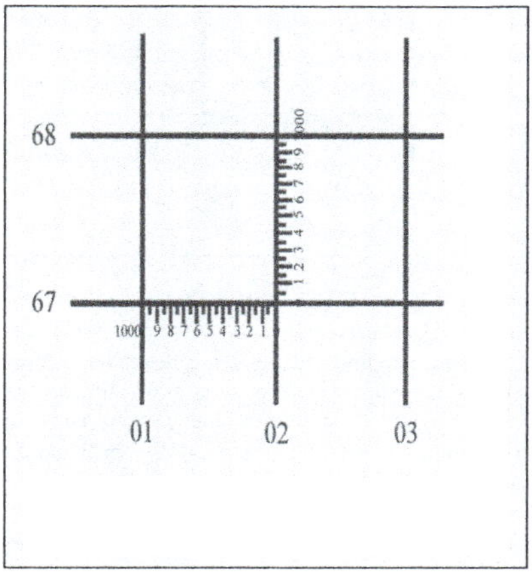

Figure 1201-5.

(c) Slide the protractor to the right until vertical grid line 02 intersects the horizontal scale at the 100-meter reading "5" (see figure 1201-6). This point 025, is your RIGHT reading.

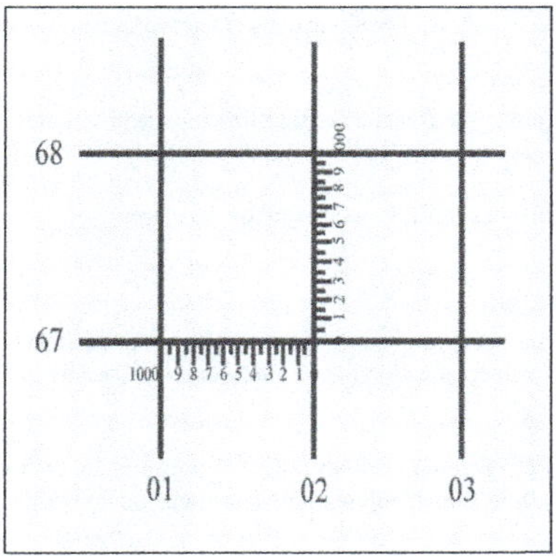

Figure 1201-6.

(d) Determine the position of your UP reading by plotting a point adjacent to your vertical scale equal to your UP reading. In this case the UP value of the vertical scale is at the 100-meter reading 2 (see figure 1201-7). This point is not only your UP reading, 672, but is also the location of your 6-digit coordinate, 025672.

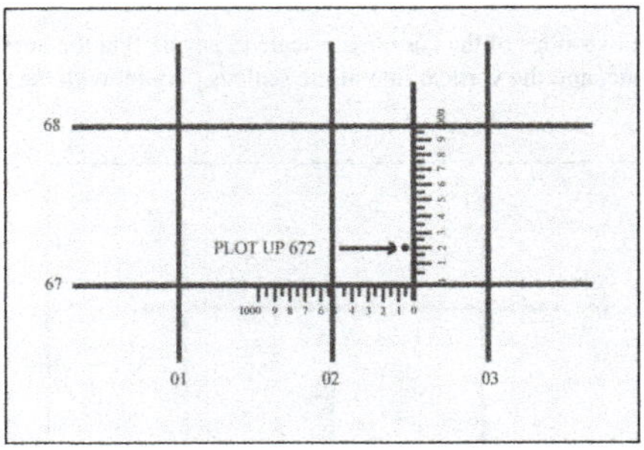

Figure 1201-7

d. 8-Digit Grid Coordinates. In some mapping situations, it is desirable to further divide a 100-meter grid square into 10-meter grid squares. This is done in the same manner as dividing a 1,000 meter grid square into 100-meter grid squares, either through estimation or by use of a coordinate scale. The result is an 8-digit grid coordinate which identifies a point on the map to within 10 meters. However, 6-digit grid coordinates, when accurately determined, should meet all your needs for effective land navigation.

4. CONTOUR LINES. A contour line is a line representing an imaginary line on the ground along which all points are the same elevation.

a. Contour lines have certain things in common.

(1) Contour lines indicate a vertical distance above or below the datum plane. (Example datum plane - sea level.)

(2) The vertical distance between adjacent contour lines is known as contour interval, and the amount of the contour interval is given in marginal information.

(3) Starting at sea level, the zero contour, each contour line represents an elevation above sea level.

(4) On most maps the contour lines indicate the nature of the slope.

b. Types Of Contour Lines.

(1) Index Contour Line. Starting at zero elevation, every fifth contour line is drawn with a heavier line. These are known as index contours. Someplace along each index contour the line is broken and its elevation is given.

(2) Intermediate Contour Lines. The contour lines falling between index contours are called intermediate contour lines. They are drawn with a finer line than the index contours and do not have their elevation given.

c. Determining Elevation. Using the contour lines on a map, the elevation of any point may be determined by:

(1) Finding the contour lines on a map from the marginal information, and noting both the amount and the unit of measure.

(2) Finding the numbered contour line nearest the point for which the elevation is being sought.

(3) Determining the direction of the slope from the numbered contour lines (index contour) to the desired point.

(4) Counting the number of contour lines that must be crossed to go from the numbered index contour line to the desired point and noting the direction - up or down. The number of lines crossed, multiplied by the contour interval is the distance above or below the starting value.

(a) If the desired point is on a contour line, its elevation is that of the contour line.

(b) For a point between contours most military needs are satisfied by estimating the elevation to an accuracy of one half the contour interval.

(c) To estimate the elevation to the top of an unmarked hill, add half the contour interval to the elevation of the higher contour line around the hill.

(d) To estimate the elevation of the bottom at a depression, subtract half the contour interval from the value of the lowest contour line around the depression.

5. TYPES OF TERRAIN FEATURES. As you plan and execute land navigation, you must be able to recognize and associate ground forms which you see or expect to see on the ground with the same features shown by contour lines on the map. Once you can recognize these ground forms and understand their characteristics, you can better use certain terrain features to your advantage and avoid those which might hinder you.

a. Hill.

(1) Representation. A hill is an area of high ground. From a hilltop, the ground slopes down in all directions. A hill is shown on a map by contour lines forming concentric circles. The inside of the smallest closed circle is the hilltop.

(2) Application To Navigation. Your ability to recognize hills on a map will help you greatly in land navigation. You can plan your routes to avoid unnecessary travel over them. Checkpoints can be chosen on or near a prominent hilltop. You can use a distant prominent hilltop to guide you. Identifying hilltops on the ground and on the map can help you locate your position accurately (see figure 1201-8).

Figure 1201-8

b. Ridge.

(1) Representation. A ridge is a series of hills that are connected to each other near the top. A ridge line may extend for many miles. It may be winding or quite straight. It may have a reasonably uniform elevation along its top or it may vary greatly in elevation.

(2) Application To Navigation. A ridge often serves as an ideal reference line during both day and night. If a map study shows that the ridge is constantly on your left, then as long as you can see the familiar characteristics of the ridge silhouetted against the skyline on your left, you know that you are heading in the proper general direction. After passing through thick terrain, if the ridge line is no longer on your left, you know you are heading in the wrong general direction. It is normally easier to maintain direction, observe steering marks, and move with less obstruction by moving along the top of the ridge as opposed to the sides of the ridge (if the tactical disadvantage of being silhouetted against the skyline is not a determining factor, and if the top of the ridge is fairly uniform) (see figure 1201-9).

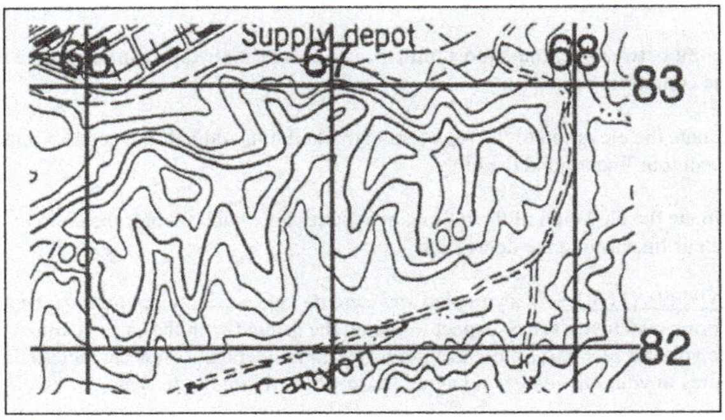

Figure 1201-9.

c. Saddle.

(1) Representation. This is a dip or low point between two areas of higher ground. A saddle is not necessarily the lower ground between two hilltops; it may be simply a dip or break along a level ridge crest. If you are in a saddle, there is high ground in two opposite directions and low ground in the other two directions. A saddle is normally represented as an hourglass or by figure-eight shaped contour lines.

(2) Application To Navigation. A distinct, unmistakable saddle will often provide a suitable checkpoint or steering mark. Also, if you must cross a ridge, it is normally easiest to cross it at a saddle (see figure 1201-10).

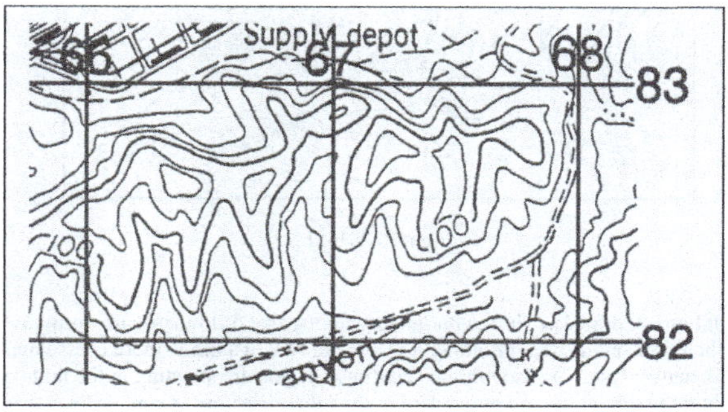

Figure 1201-10.

d. Finger.

(1) Representation. A finger is a short, continuous sloping line of higher ground, normally jutting out from the side of a ridge or hill. A finger is often formed by two roughly parallel draws. The ground slopes down in three directions and up in one. Contour lines on a map depict a finger with the U or V pointing away from high ground.

(2) Application To Navigation. A distinct, unmistakable finger may be chosen as a checkpoint or steering mark. In the defense, fingers provide good fields of view for observation posts. During movement, there is a chance of being skylined when traveling on fingers (see figure 1201-11).

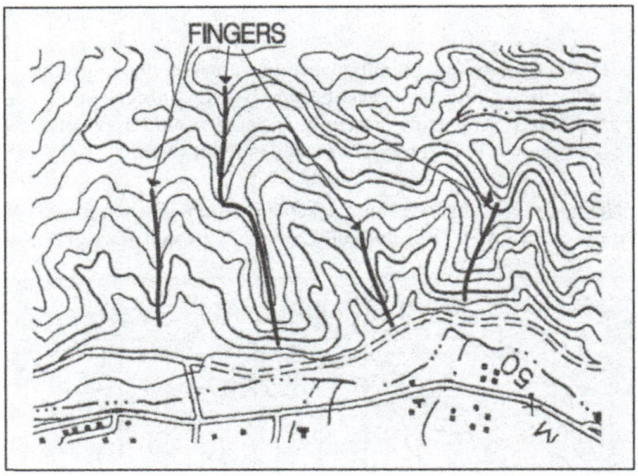

Figure 1201-11

e. Draw.

(1) Representation. A draw is a short, continuous sloping line of low ground, normally cut into the side of a ridge or hill. Often, there is a small stream running down the draw. In a draw, there is essentially no level ground. Therefore, little or no maneuver room exists within its confines. If you are standing in the middle of a draw, the ground slopes upward in three directions and downward in the other direction. Contour lines (see figure 1201-12) on a map depict a draw with the U or V pointing toward high ground.

(2) Application To Navigation. A draw may or may not be a wise route to follow. The sides of a draw are often steep. The sides of draws are normally steeper than the fingers on both sides of the draw. Fingers normally have convex slopes while draws normally have concave slopes. Therefore, when you must climb a hill, draws provide an easy route at the bottom of the hill, but the route gets more difficult near the top when the sides of draws are normally quite steep. Draws often provide good concealment during movement, but if the draw is covered by an enemy machine gun, it could become a death trap due to its steep sides. A draw can be used to maintain direction.

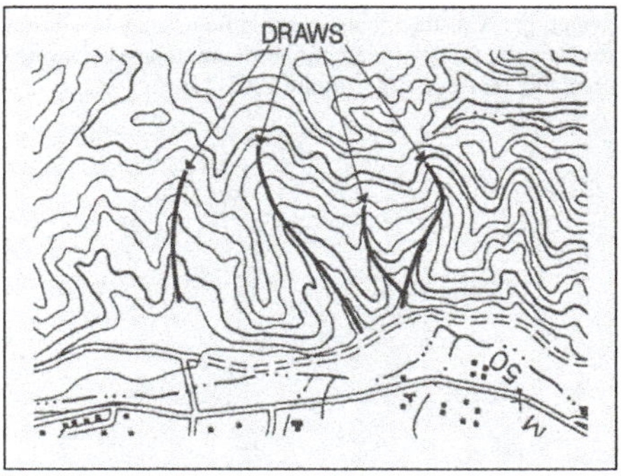

Figure 1201-12

(3) Distinguishing Between Draws And Fingers. Look at point "A" on figure 1201-13. Is it a finger or a draw? If you think it is a finger, you are wrong. If you think it is a draw, you also are wrong. You see, from the

amount of information given in the figure, you can not tell what type of terrain feature it is. You MUST look at the surrounding terrain features. If there is a hill in the direction of the open end of the "U" then the feature is a finger; if not, then it is a draw. If there is a stream in the middle of the feature, then is it a finger or a draw? That's right, a draw. Along those lines, here's a trivia question for you - which way does the closed end of the "U" of the contour lines point on a stream, upstream or downstream? You already have enough information to answer this. Which way does the closed end of the "U" of the contour lines point in a draw? And where are streams, in draws or fingers? So the correct answer is that the closed end of the "U" points UPSTREAM.

Figure 1201-13

f. Valley.

(1) Representation. A valley is continuous line of low ground that normally rests in between two roughly parallel ridges. Valleys often have streams or rivers running through them.

(2) Application To Navigation. If you are navigating through a valley that has a stream, keep in mind that while a dry stream bed provides a relatively vegetation-free, easy-to-follow route, it will often meander back and forth across the valley floor. During inclement weather, many streams are subject to flash flooding. Additionally, the banks of streams, creeks, and rivers often contain mud flats, marshes, and extremely thick vegetation. During tactical foot movements, keep in mind the need for security and silent movement. Moving silently while wading in water requires a great deal of effort on the part of each Marine. Therefore, while the trace of a stream will often provide an ideal reference line, it is recommended that you move away from the stream and use the valley as a reference point.

g. Cuts And Fills.

(1) Representation.

(a) A CUT is a man-made feature resulting from cutting through high ground, usually to form a level bed for a road or railroad track. Cuts are drawn with a contour line along the cut line. This contour line extends the length of the cut and has tick marks that extend from the cut line to the roadbed, if the map scale permits this level of detail (see figure 1201-14).

(b) A FILL is a man-made feature resulting from filling a low area, usually to form a level bed for a road or railroad track. Fills are drawn with contour lines along the fill line. This contour line extends the length of the filled area and has tick marks that point toward lower ground. If the map scale permits, the length of the fill tick marks are drawn to scale and extend from the base line of the fill (see figure 1201-14).

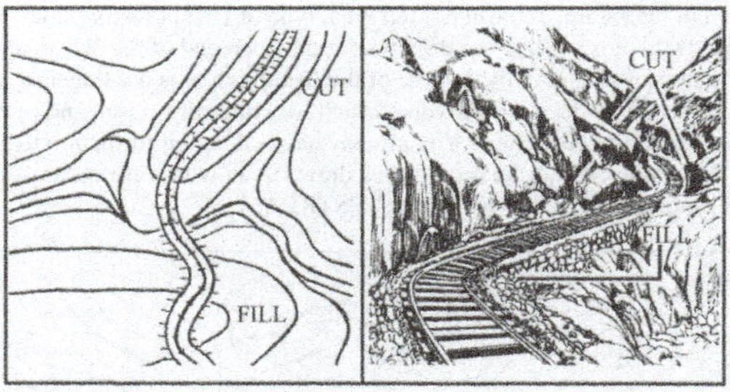

Figure 1201-14

(2) Application To Navigation. Prominent cuts and fills can often be used as checkpoints. If the map shows that a road or railroad is punctuated with cuts and fills, it indicates that efforts have been made to eliminate radical changes in slope.

h. Depression.

(1) Representation. This is a low point in the ground or a sinkhole. It is an area of low ground surrounded by higher ground in all directions, or simply a hole in the ground. Usually only depressions that are equal to or greater than the contour interval will be shown. On maps, depressions are represented by closed contour lines that have tick marks pointing toward low ground (see figure 1201-15).

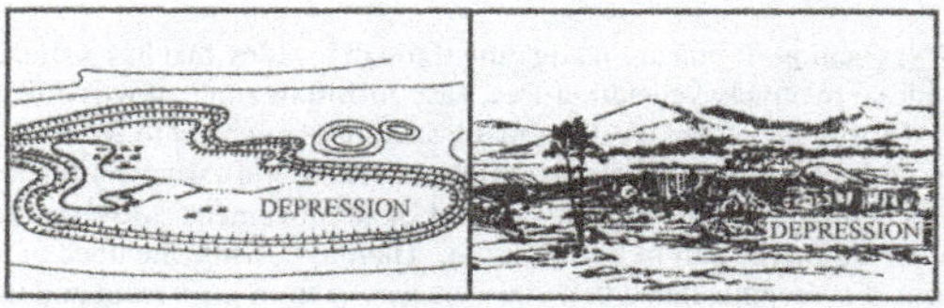

Figure 1201-15

(2) Application To Navigation. Prominent depressions can often be used as checkpoints. Because of the usually steep sides of depressions, bypass them if possible.

i. Cliff.

(1) Representation. A cliff is a vertical or near vertical terrain feature. It is an abrupt change of the land. When a slope is so steep that the contour lines converge into one "carrying" contour of contours, this last contour line sometimes has tick marks pointing toward low ground (see figure 1201-16).

(2) Application To Navigation: Prominent cliffs can often be used as checkpoints. Because of the steep face of a cliff, it should be bypassed.

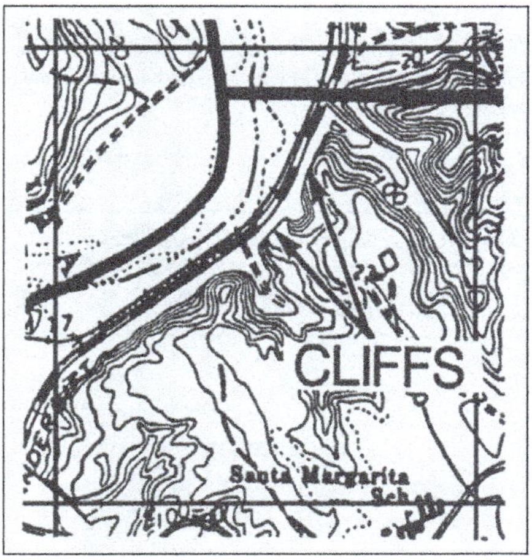

Figure 1201-16

6. DETERMINE DISTANCE ON A MAP.

a. Determining Straight-Line Distance.

(1) To determine straight-line distance between two points on a map, lay a straight-edged piece of paper on the map so that the edge of paper touches both points and extends past them. Make a tick mark on the edge of the paper at each point (see figure 1201-17). Remember that the center of the topographic symbol accurately designates the true location of the object on the ground; therefore, measure all map distances from the center of the topographic symbol.

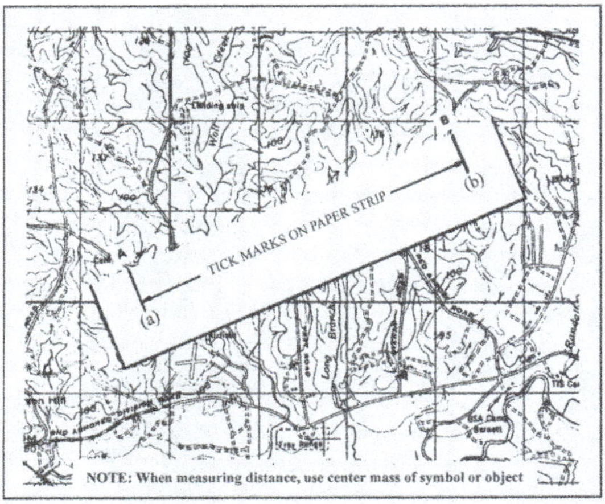

Figure 1201-17

(2) To convert map distance to ground distance, move the paper down to the appropriate unit of measure on the graphic bar scale, and align the right tick mark (b) with a printed number in the primary scale so that the left tick mark (a) is in the extension scale (figure 1201-18).

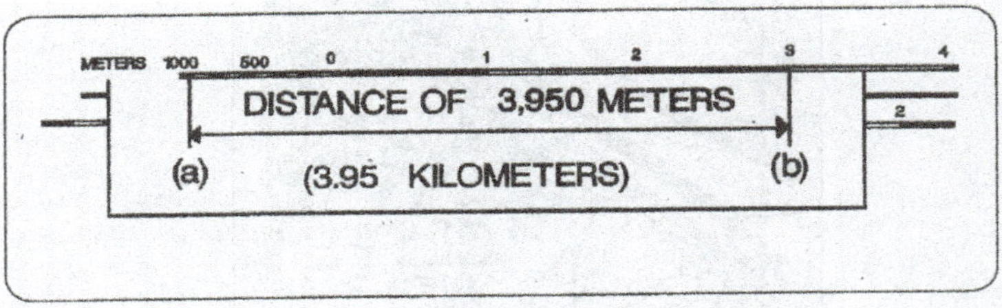

Figure 1201-18

In this case the right tick mark (b) is aligned with the 3,000 meter mark in the primary scale, thus the distance is at least 3,000 meters. To determine the distance between the two points to the nearest 10 meters, look at the extension scale. The extension scale is numbered with zero at the right and increases to the left. When using the extension scale, always read RIGHT TO LEFT (see figure 1201-18). From the zero to the end of the first shaded square is 100 meters. From the beginning of the white square to the left is 100 to 200 meters; at the beginning of the second shaded square is 200 to 300 meters. Remember, the distance in the extension scale increases from right to left. To determine the distance from tick mark (a), estimate the distance inside the squares to the closest tenth. As you break down the distance between the squares in the extension scale, you will see that tick mark (a) is aligned with the 950-meter mark. Adding the distance of 3,000 meters determined in the primary scale, we find that the total distance between (a) and (b) is 3,000 + 950 = 3,950 meters.

(3) There may be times when the distance you measure on the edge of the paper exceeds the graphic scale. One technique you can use to determine the distance is to align the right tick mark (b) with a printed number in the primary scale, in this case 5 kilometers (see figure 1201-19). You can see that from point (a) to (b) is more than 6,000 meters. To determine the distance to the nearest 10 meters, place a tick mark (c) on the edge of the paper at the end of the extension scale (see figure 1201-19).

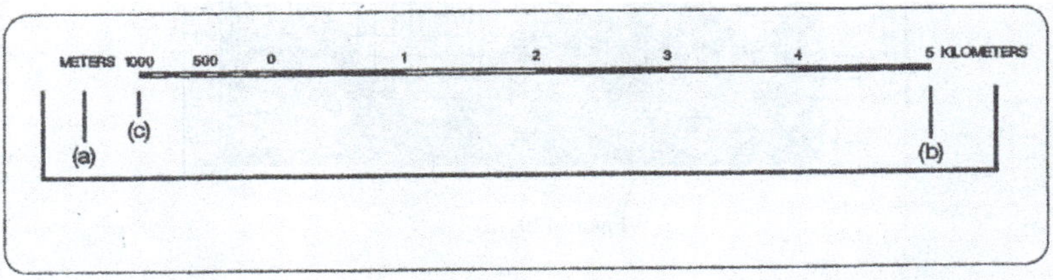

Figure 1201-19

You know that from point (b) to (c) is 6,000 meters (5,000 from the primary scale and 1,000 from the extension scale). Now, measure the distance between points (a) and (c) on your sheet of paper the same way you did earlier, only use point (c) as your right hand tick mark (see figure 1201-20). The total ground distance between start and finish points is 6,420 meters.

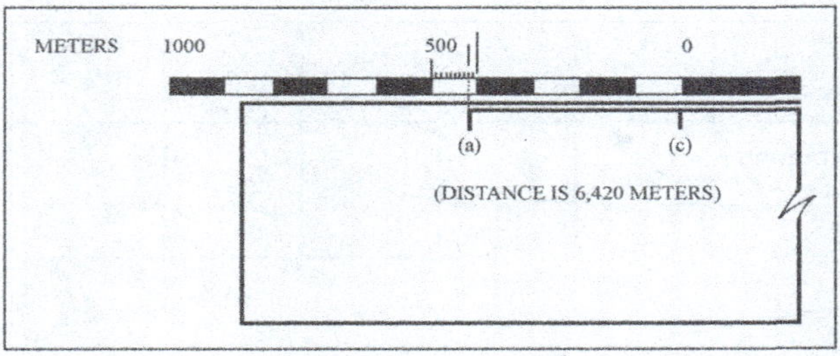

Figure 1201-20

(4) One point to remember is that distance measured on a map does not take into consideration the rise and fall of the land. All distances measured by using the map and graphic scales are flat distances. Therefore, the distance measured on a map will increase when actually measured out on the ground (see figure 1201-21).

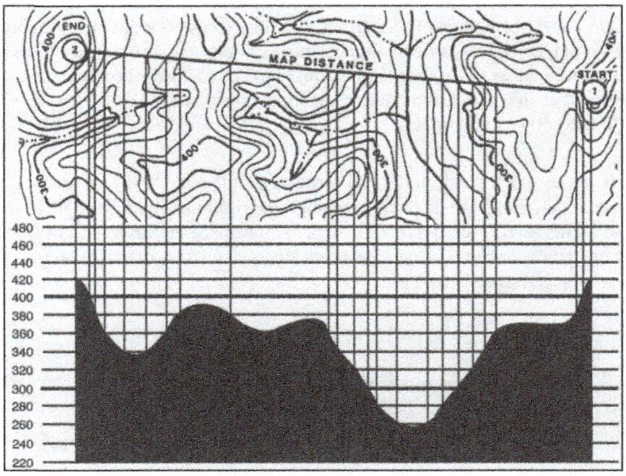

Figure 1201-21.

b. Determining Irregular Map Distance. To measure distance along a winding road, stream, or other curved line, you still use the straight edge of a piece of paper. In order to avoid confusion concerning the starting point and the ending point, a six-digit coordinate, combined with a description of the topographical feature, should be given for both the starting and ending points. Place a tick mark on the paper and map at the beginning point from which the curved line is to be measured. Place a paper strip or other material with a straightedge along the center of the irregular feature (see figure 1201-22), and extend the tick mark onto the paper strip. Because the paper strip is straight and the irregular feature is curved, the straightedge will eventually leave the center of the irregular feature. At the exact point where this occurs, place a tick mark on both the map and paper strip.

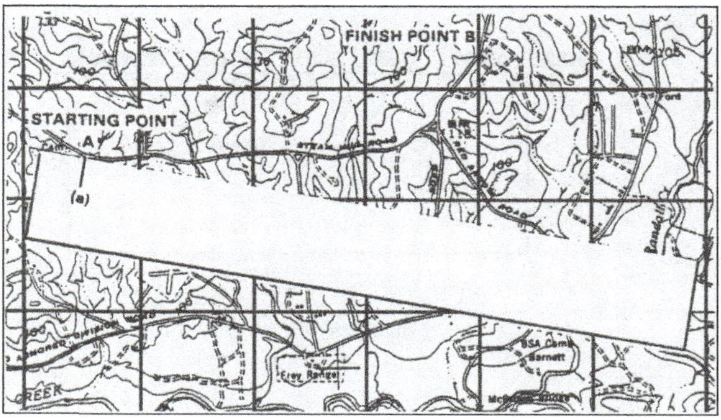

Figure 1201-22

Keeping both tick marks together (on paper and map), place the point of the pencil close to the edge of the paper on the tick mark to hold it in place and pivot the paper until another straight portion of the curved line is aligned with the edge of the paper. Repeat this procedure while carefully aligning the straightedge with the center of the feature and placing tick marks on both the map and paper strip each time it leaves the center until you have ticked off the desired distance (see figure 1201-23).

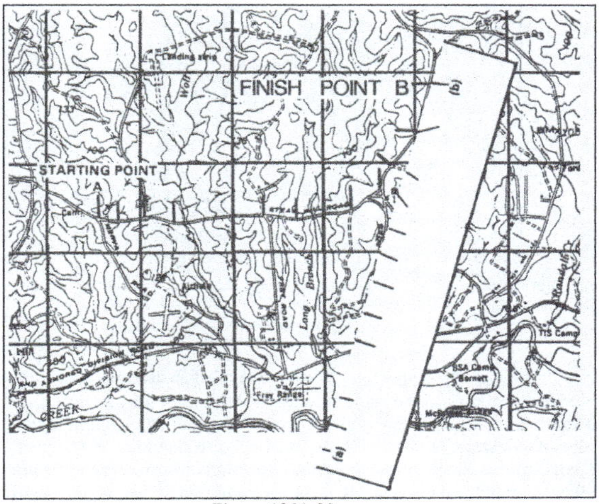

Figure 1201-23.

Place the paper strip on a graphic bar scale and determine the ground distance measured.

7. PLOTTING A GRID AZIMUTH.

a. Azimuth Defined. An azimuth is a straight line from one point to another. This definition, however, leaves much to be desired in understanding the complete meaning of the word azimuth as it applies to land navigation. Therefore, the following definition is required: An azimuth is an angle measured in a clockwise direction from a predetermined base line. Before attempting to determine or follow an azimuth in the field, you must have a clear understanding of each part of this definition.

(1) An Azimuth Is An Angle. "An azimuth is an angle," means that it is part of a circle. But just how much of a circle does each azimuth represent? If a circle were divided into 360 equal "slices of pie" each slice would be one degree (see figure 1201-24).

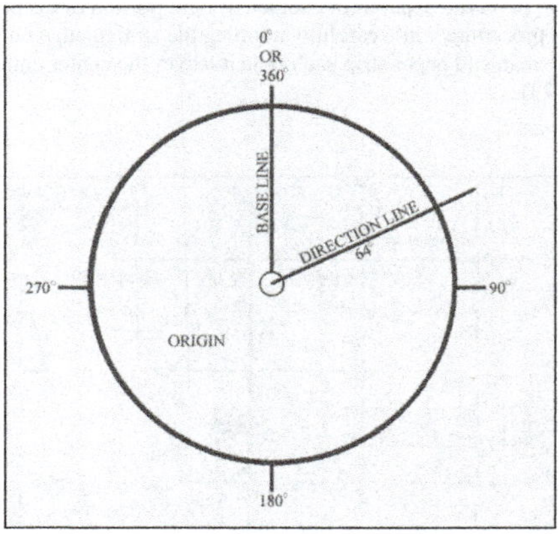

Figure 1201-24

(2) Measured In A Clockwise Direction. When we say that an azimuth is an angle measured in a clockwise direction, we mean that each of the angles discussed above must have a starting point and from there progress in numerical value in a clockwise direction around the circle until they return to the starting point. The starting point has a value of 0 degrees, or (since it is also the final direction line) 360 degrees. It may be expressed as either 0 degrees or 360 degrees. Since the degree value of azimuths always progresses in a clockwise direction, all azimuths between 0 degrees and 180 degrees will be on the right side of the imaginary circle and all azimuths between 180 degrees and 360 degrees will be on the left side of the circle (see figure 1201-24). Keep in mind that there are only 360 degrees in a circle. When working map problems, if you mathematically arrive at a figure exceeding 360 degrees, then you have gone completely around the circle and started over again. For example, if you add 15° to 350°, it would be expressed as 5°, NOT 365°.

(3) From A Predetermined Base Line. When we say that an azimuth is an angle measured in a clockwise direction from a predetermined base line, we mean that it is a certain number of degrees measured in a clockwise direction from some sort of reference point. It is this portion of the definition that causes the most misunderstanding, confusion, and often loss of direction in the field. This reference point or base line we are referring to is north. There are three base lines--true north, magnetic north, and grid north (see figure 1201-25). The most commonly used are magnetic and grid north.

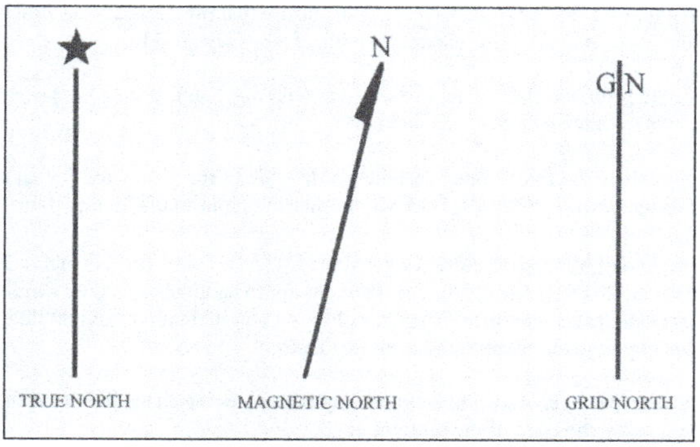

Figure 1201-25

(a) True North. The true north line is a line from any point on the earth's surface to the North Pole. True north can be found at night by locating the North Star, which always points towards true north. True north is usually represented on the declination diagram by a line ending with a star (see figure 1201-25). True north is used almost exclusively when navigating without a compass.

(b) Magnetic North. The earth has a magnetic field that is close to (but not exactly on) the North Pole. The direction to this north magnetic pole is indicated by the north-seeking arrow of your lensatic compass. Magnetic north is usually symbolized on the declination diagram by a line ending with a half arrowhead (see figure 1201-25). Anytime you use the compass to plan or follow an azimuth in the field, you must work with azimuths measured from magnetic north.

(c) Grid North. This base line is established by using the vertical grid lines on the map. Grid north may be symbolized on the declination diagram by the letters GN (see figure 1201-25). Anytime you use a protractor in conjunction with a vertical grid line to determine or plot an azimuth on a map, you must work with an azimuth measured from grid north.

(4) When using grid north and magnetic north as base lines, you need to understand the difference between a grid azimuth and a magnetic azimuth:

(a) A grid azimuth is an angle measured in a clockwise direction from grid north.

(b) A magnetic azimuth is an angle measured in a clockwise direction from magnetic north.

b. Determining A Grid Azimuth. There are two methods of measuring grid azimuths from one point to another on the map. Whichever method you use, remember that you are dealing with GRID AZIMUTHS. A grid azimuth CAN NOT be followed with a compass.

(1) Protractor And String Method. To use this method, you must first modify your protractor. Using a needle and piece of thread, punch a SMALL hole through the index mark of your protractor. (The thread should be about 6 inches long.) Tie a knot in the thread on each side of the protractor as close to the index line as possible to secure the thread to the protractor. Now you are ready to go.

(a) Place the index mark on your starting point. Ensure that the vertical base line is parallel with a north-south grid line and the horizontal base line is parallel with an east-west grid line.

(b) Holding the protractor firmly against the map with one hand, stretch the piece of thread with your other hand so that the thread intersects your second point.

(c) The point where the thread intersects the inside scale of the protractor is your azimuth (the outside scale is in mils). This is the GRID AZIMUTH from the starting point to the second point.

(2) Protractor and Pencil Method. Before we get started with the steps for this method, you need to understand something. If you go in a STRAIGHT line from one point to another, does the azimuth ever change? Of course not. No matter where you are on that line of march, as long as you maintain a straight line, the azimuth never changes. Remember that when you get to Steps 1 and 3 for this method.

(a) Using a straight edge, draw a line connecting the two points. This line needs to be at least 4 inches long so that the line reaches the edge of the protractor.

(b) Label the two points A and B, with your point A being your starting position.

(c) Place the index mark of your protractor on the line you just drew where that line intersects a vertical grid line. This point should be as close to point A as possible. Ensure that the entire vertical base line on your protractor is directly on the vertical grid line that your line intersects. If it is not, then your azimuth reading will be inaccurate.

NOTE: The reason for placing the index mark on a point where your line intersects a vertical (or horizontal) grid line instead of directly on point A is so that you get a more accurate reading. Remember, as long as you are on that straight line from point A to point B, the azimuth does not change.

(d) The point where the line intersects the inside scale of the protractor is your azimuth (the outside scale is in mils). This is the GRID AZIMUTH from point A to point B.

c. Plotting A Grid Azimuth. To plot a grid azimuth on a map, follow these steps:

(1) Place the protractor on the map with the index mark at center mass of the known point.

(2) Ensure that the protractor's vertical base line is parallel with the closest north-south grid line and the horizontal base line is parallel with an east-west grid line.

(3) Make a mark on the map at the desired grid azimuth.

(4) Remove the protractor and draw a line connecting the known point and the mark on the map. You have now plotted the grid azimuth.

(5) To plot a known distance onto that azimuth, use a piece of paper to transfer the distance from the scale to the plotted azimuth. Remember, the azimuth you have just plotted is a GRID AZIMUTH. It can not be followed using a compass.

d. Determining A Back Azimuth. A back azimuth is the opposite direction of an azimuth. It is the same as doing an "about face." To obtain a back azimuth from an azimuth, ADD 180 degrees if the azimuth is 180 degrees or less or subtract 180 degrees if the azimuth is 180 degrees or more. The back azimuth of 180 degrees may be stated as 0 degrees or 360 degrees. So, what happens if you add when you should have subtracted or subtracted when you should have added? You will come up with either a negative number, or one that is over 360 degrees. Since you cannot express azimuths as either negative numbers or as numbers over 360 degrees then simply do the opposite of what you did before. If you added 180 degrees, then subtract. If you subtracted 180 degrees, then add.

8. NAVIGATE USING THE LENSATIC COMPASS.

a. Lensatic Compass Description. The lensatic compass consists of three major parts:

- cover
- base
- rear sight

(1) Cover. This protects the floating dial and the glass encasement. It contains the sighting wire and two luminous sighting dots for night navigation.

(2) Base.

(a) Floating Dial. This is mounted on a pivot so that it rotates freely when the compass is held level. It contains the magnetic needle. A luminous arrow and the letters "E" and "W" are printed on the dial. The arrow points to magnetic north. Letters fall at the east (E) 90 degrees and (W) 270 degrees. There are two scales: outer - denotes MILS (black); inner - denotes DEGREES (red).

NOTE: Mil, is another unit of measure. The mil (abbreviated m), is mainly used in artillery, tank, and mortar gunnery. The mil expresses the size of an angle formed when a circle is divided into 6,400 angles with the vertex of the angles at the center of the circle. A relationship can be established between degrees and mils. A circle equals 6400 mil divided by 360 degrees, or 17.78 mils. To convert degrees to mils, multiply degrees by 17.78.

(a) Glass Encasement. This houses the floating dial and contains a fixed black index line.

(b) Bezel Ring. This device that clicks when turned. It contains 120 clicks when rotated fully. Each click equals 3 degrees. A short luminous line is used in conjunction with the north-seeking arrow during night navigation.

(c) Thumb Loop. This is attached to the base.

(3) Rear Sight. This is used to lock the floating dial. The rear sight must be opened more than 45 degrees to allow the floating dial to float freely.

(a) Lens. This is used to read the floating dial.

(b) Rear Sight Slot: This is used in conjunction with the front sighting wire when aiming at objects.

b. Handling The Compass.

(1) Inspection. Compasses are delicate instruments and should be cared for accordingly. Conduct a detailed inspection before you use your compass. The most important part is the floating dial. It must float freely. You must also ensure the sighting wire is straight, the glass encasement is not broken, and the numbers on the dial are readable.

(2) Effects Of Metal And Electricity. Metal objects and electrical sources can affect the performance of a compass (non-magnetic metals and alloys do not affect compass readings). To ensure the proper functioning of your compass, keep a safe distance from the following metal objects:

High-tension power lines	55 meters
Field gun, truck, or tank	10 meters
Telegraph or telephone wires and barbed wire	10 meters
Machine gun	2 meters
Rifle	1/2 meter
Steel rim glasses	1/3 meter

(3) Accuracy. A compass in good working condition is very accurate. However, you must periodically check your compass on a known line of direction, such as a surveyed azimuth using a declination station. If your compass has more than 3 degrees ± variation, do not use it.

(4) Protection. When traveling with the compass unfolded, make sure the rear sight is folded down onto the bezel ring. This will lock the floating dial and prevent vibration, as well as protect the crystal and rear sight from being damaged.

b. Techniques For Using The Compass. The lensatic compass is used to determine or follow magnetic azimuths during both the day and night. To use it with the maximum degree of accuracy, it is important that certain techniques be understood and properly applied. Like developing techniques for shooting a rifle, you must develop the proper holding position and practice until you master the techniques for accurately "shooting" an azimuth.

(1) Using The Centerhold Technique. This technique is faster and easier to use than the other techniques (see figure 1201-26).

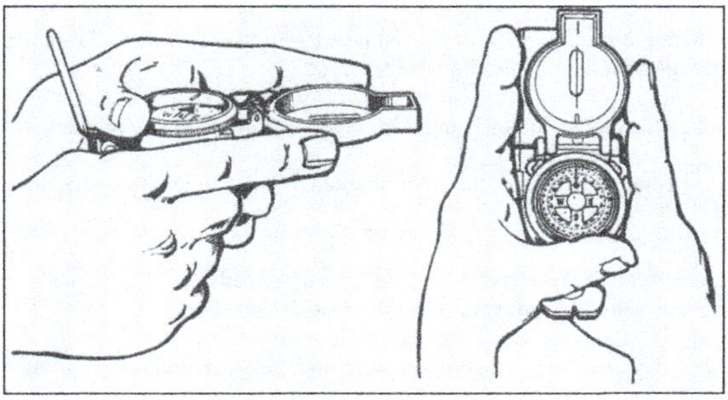

Figure 1201-26.

(a) Open the compass cover so that the cover forms a straightedge with the base.

(b) Move the rear sight to the rearmost position to allow the dial to float freely.

(c) Place your thumb through the thumb loop and form a steady base with your third and fourth fingers. Extend both index fingers along the sides of the compass.

(d) Place the thumb of the other hand between the lens (rear sight) and the bezel ring; place the remaining fingers around the fingers of the other hand.

(e) Pull your elbows firmly into your sides.

(f) Turn your entire body until the desired magnetic azimuth is under the fixed black index line.

(2) Using The Compass-To-Cheek Technique. This technique is used when you want to be more accurate (see figure 1201-27). Follow these steps.

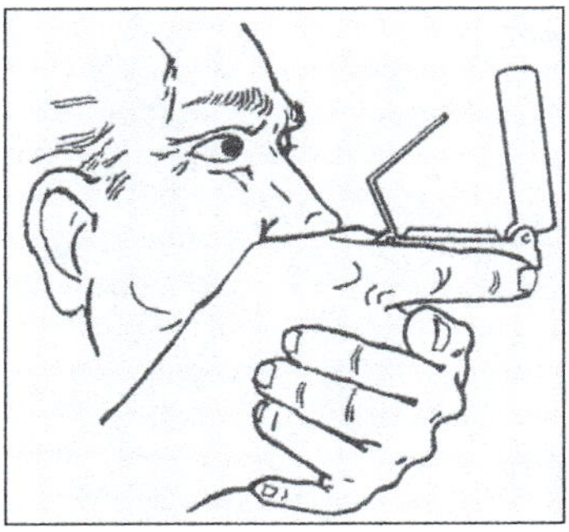

Figure 1201-27

(a) Open the compass so that the cover is vertical, forming a 90 degree angle with the base.

(b) Move the rear sight to the rearmost position to release the dial, then fold it slightly forward.

(c) Turn the thumb loop all the way down and insert your thumb. Form a loose fist under the compass, steady it with your other hand, and raise it to eye level.

(d) Look through the rear sight notch and center the front sighting wire in the rear sight notch.

(e) Keeping the compass level and the sights aligned, rotate your entire body until the sighting wire is lined up on a distant object.

(f) Glance down through the lens and read the azimuth directly under the black index line. The azimuth you read is the magnetic azimuth from your position to the distant object.

c. Converting Azimuths. Because there is an angular difference between grid north and magnetic north, a conversion from magnetic to grid or vice versa is needed.

(1) Declination Diagram. As previously discussed, azimuths measured with a protractor are grid azimuths (measured from grid north), and azimuths determined with the compass are magnetic azimuths (measured from magnetic north). You cannot follow a grid azimuth with a compass, nor can you plot a magnetic azimuth with a protractor because of the angular difference between grid north and magnetic north. This angular difference (between grid north and magnetic north) is called the G-M ANGLE (Grid-Magnetic angle**). The G-M angle varies for each map.** Because of this angular difference (the G-M angle), before you can plot a magnetic azimuth on a map, you must convert it to a grid azimuth. Likewise, before you can use a grid azimuth to navigate, you must convert it to a magnetic azimuth. Declination diagrams display the difference between grid and magnetic north. A complete set of instructions is included in MOST declination diagrams for your use in converting azimuths. The G-M angle often is not expressed as a whole degree, such as 1/2 degrees or 7 degrees 15'. Since you will not need to work with such precise numbers as minutes, round the G-M angle off to the nearest whole degree. If the G-M angle is 1/2o or 30' then round the angle up to the next highest whole degree.

(2) Conversion Notes. Refer to the conversion notes that appear with the declination diagrams explaining the use of the G-M angle (see figure 1201-28). One note provides instructions for converting a magnetic azimuth to a grid azimuth. The other provides instructions for converting a grid azimuth to a magnetic azimuth. The conversion (addition or subtraction) is governed by the direction of magnetic north relative to grid north.

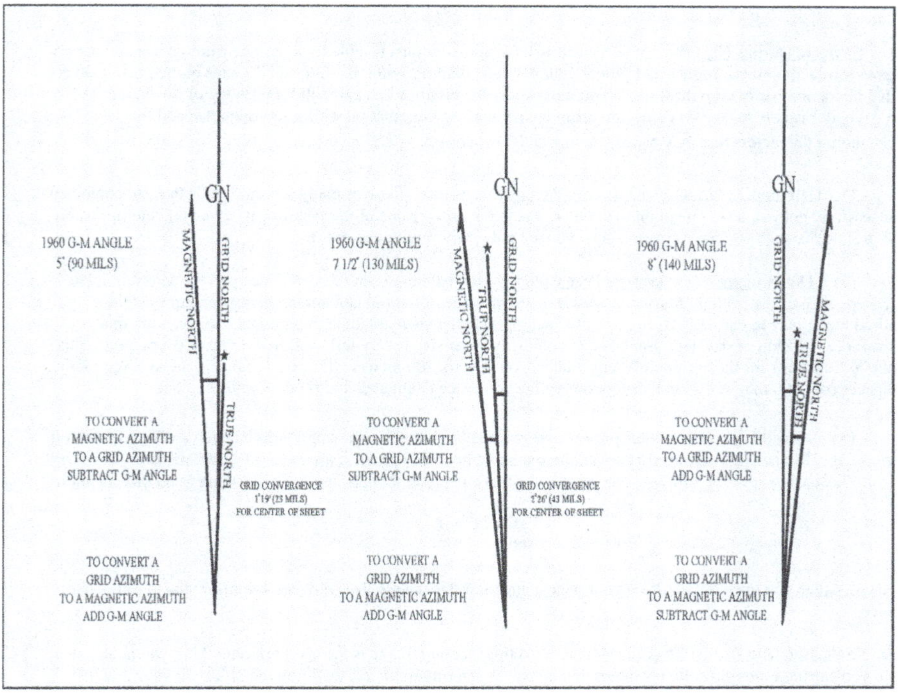

Figure 1201-28

(a) Converting Grid Azimuths To Magnetic Azimuths. The following is an example of part of a declination diagram:

TO CONVERT A
GRID AZIMUTH TO A
MAGNETIC AZIMUTH
SUBTRACT G-M ANGLE

On a map with a G-M angle of 15 degrees, to convert a 39 degrees grid azimuth to a magnetic azimuth you simply follow the instructions. Subtracting the G-M angle (15 degrees) from the grid azimuth (39 degrees), you get the correct magnetic azimuth of 24 degrees.

(b) Converting Magnetic Azimuths To Grid Azimuths. Again, examine the following declination diagram. The conversion note states:

TO CONVERT A
MAGNETIC AZIMUTH
TO A GRID AZIMUTH
ADD G-M ANGLE

To convert a 238 degrees magnetic azimuth to a grid azimuth you simply follow the instructions. Adding the G-M angle (15 degrees) to the magnetic azimuth (238 degrees), gives you the correct grid azimuth of 253 degrees.

d. Navigate During The Day. To successfully navigate, you must be able to determine from your map the best routes to follow, plot these routes, and follow your desired compass azimuth. Almost all routes plotted on a map are divided into a number of straight "legs." You must know the distance and azimuth for each leg of the march. A good navigator is able to stay on course by using the compass in conjunction with steering marks, and by understanding the factors that may cause you to wander off course.

(1) Determining Ground Distance By Pacing. Navigating involves many separate skills that are combined to successfully get you from one point to another. Pacing is another one of these skills. It helps you determine how much ground distance you have covered once you actually start moving on your route.

(2) Determining Your Average Pace Count. Many techniques have been developed for determining how many steps it takes the average Marine to walk a given distance. Predominant among these techniques is the "hundred meter pacing course." A distance of exactly 100 meters is marked on the ground. As you walk this distance, keep a count of the steps you take. Count every other step (every left foot) to determine your count. This is your "pace count" for the number of steps it takes you to walk 100 meters. The pace count for the average Marine is 60 paces (per 100 meters). If you cannot determine your pace count ahead of time, use this number.

(3) Now that you know your pace count, you need to know how to apply it to land navigation. Once you determine the distance that you need to cover (for example, how far you must travel on a leg of a patrol), you must convert this distance into the number of steps you must take to cover that distance. The formula to do this is simple:

$$\frac{D}{100} \times PC = P$$

D is the distance you must travel, PC is your pace count, and P is the number of paces you must take to travel that distance.

e. Selection And Use Of Steering Marks. Selecting steering marks is the last important thing you must know before you can start moving from one point to another. A steering mark is a well-defined object on your line of march on which you can guide. These objects can be natural or man-made (hill, tree, building, etc.), a celestial body (sun, stars, moon), or another person. One of the problems associated with selecting and using steering marks is that an object often looks good when you select it, but will become obscured as you approach it. This may confuse you and cause you to deviate from your intended line of march. The characteristics of good steering marks are discussed below and should be kept in mind when you select a steering mark.

(1) A good steering mark must have some distinct and unique features such as:

- Color
- Size
- Shape

A good steering mark will have all three. This assures you that it will continue to be recognizable as you approach it. A distant tree may appear to have a crooked limb which identifies it. However, upon entering the forest you may find dozens of widely separated trees with similar limbs. Your steering mark has then served little purpose. A steering mark must have a feature that distinguishes it from all other similar objects on your line of march.

(2) If several easily distinguishable objects appear along your line of march, the best steering mark would be the most distant object and/or the highest. This will enable you to travel farther with fewer references to the compass. The higher steering mark is not as easily lost to sight as is a low steering mark that may blend into the background as you approach it.

(3) A steering mark should be continuously visible. If the terrain or vegetation ever blocks the steering mark from view, take out your compass and select an intermediate steering mark. Continue using intermediate

steering marks until your original steering mark comes back into view. To enter a wooded area thinking that you will come out on the other side still heading in the correct direction is foolish. In thick vegetation, you can quickly become disoriented and emerge from the thicket heading the wrong way with no steering mark in view.

(4) If you are navigating alone and there are no usable steering marks to your front, it is still possible to proceed on an azimuth by referring to a back steering mark. Simply determine the back azimuth of the azimuth you are following. Now face about, and see if there is a prominent object on line with the back azimuth. If none exist, erect one where you are located (place a stick in the ground, pile up some rocks, etc.) When you continue on your route, use the same principles that apply for a forward steering mark, except that you use back azimuths to maintain the intended line of march and you must check your compass more frequently.

(5) If appropriate landmarks are not available at night, you may select a plainly visible star along your line of march to serve as a steering mark. Remember though, due to the Earth's rotation, any star that you choose will eventually either disappear under the horizon or will move too high in the sky to be of further use. If this happens, choose another steering mark.

(6) If no natural or man-made object is visible on your line of march, have another Marine move forward to serve as a steering mark. The distance he moves is determined by the terrain, amount of visibility, and the tactical situation. Since this is time-consuming and may be tactically compromising, you should never use this technique except as a last resort.

(7) Steering marks are selected as the march progresses. Shoot your azimuth, select the best steering mark on this azimuth, and head to it. Whether you are navigating during daylight or periods of reduced visibility, through densely wooded areas or open terrain, over short or great distances, every step you take should be toward a selected steering mark. The natural human tendency to veer off course, even on short distances, is too great to needlessly trust your ability to walk a straight line. So, whether it is an object which meets all the requirements of an ideal steering mark, a star, or something as insignificant as a tiny patch of light in the foliage, ALWAYS, choose something on your line of march.

f. Following A Compass Azimuth During The Day. The procedure for following a compass azimuth during the day utilizes the compass-to-cheek technique and the selection and use of appropriate steering marks. Let's begin.

(1) Shoot your desired azimuth using the compass-to-cheek technique.

(2) Using the sighting slot and the sighting wire, choose the best steering mark that is directly in line with your azimuth.

(3) Recheck your azimuth. The few seconds required for this are well spent.

(4) Close your compass to protect it during movement and step off towards your steering mark.

(5) Periodically spot check your azimuth by using the centerhold technique. You can do this without stopping.

(6) When you arrive at your steering mark, stop and select a new steering mark. Continue repeating the steps until your journey is complete.

NOTE. If your steering mark becomes hidden from your view, stop and select another steering mark immediately.

g. Navigating Using Dead Reckoning. The term "dead reckoning" derives from the seafaring days before the arrival of modern navigational equipment when mariners would "deduce" (ded) how far they had traveled on their course. Based on this deduction, they would "reckon" where they were located. To save room in the ship's log, entries were abbreviated to read "ded-reckon." "Dead-reckoning" is a means of navigating using only pace counts and azimuths. Terrain association is not used. In thick jungles, barren terrain, or when visibility is limited, you will not be able to associate your current position with visible terrain features on the ground. Dead reckoning may be the only means you have available of knowing where you are located on your map. To navigate by dead reckoning, you must accomplish the following:

(1) Determine Location, Distance And Azimuth. Determine your own location by any means. Then determine the distance and magnetic azimuth between each leg of your journey.

(2) Step Off. During movement, make frequent reference to your compass to ensure you are moving in the correct direction. Carefully select your steering marks (navigating with steering marks is NOT the same as navigating by terrain association as steering marks are rarely shown on a map). Maintain an accurate pace count. When you have paced off the appropriate number of steps on the appropriate azimuth, then you have reached your objective (or checkpoint). Repeat the steps until you reach your final destination.

h. Following a Compass Azimuth at Night. The procedure for following an azimuth at night is basically the same as for following an azimuth during daylight. In fact, on a bright, moonlit night, the procedures are identical since you can see the compass dial clearly. However, because of the reduced visibility, night navigation will require further consideration in selecting steering marks and applying the features of the lensatic compass intended for night use.

(1) Setting The Compass For Night Use. The lensatic compass is equipped with special features which enable you to follow an azimuth during periods of darkness. Around the base of the bezel ring is a series of 120 notches. On the forward edge of the body of the compass is a tiny bezel detent spring with its tip seated in one of the notches. As the bezel ring is turned, the spring moves from notch-to-notch producing "clicks." Each click equals 3°. This clicking action, used in conjunction with the luminous markings on the compass, provides a means of setting your compass at night.

(2) Using The Click Technique. At night, an azimuth can be set on the compass by using the click technique (see figure 1201-29). Remember that the bezel ring contains 120 clicks; each click equals 3 degrees. Use the following steps to set a compass for night navigation.

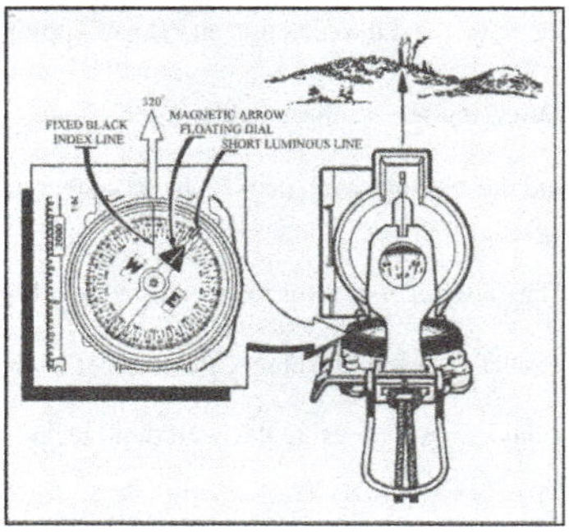

Figure 1201-29

(a) Rotate the bezel ring until the short luminous line is directly over the fixed black index line.

(b) Divide the azimuth you plan to travel by 3 to get the number of clicks you must rotate the bezel ring.

(c) Rotate the bezel ring the desired number of clicks in a counterclockwise direction.

NOTE: If the number of clicks is greater than 60, you can subtract this from 120 and rotate the bezel ring in a clockwise direction. For example, you would move the bezel ring 20 clicks clockwise instead of 100 clicks counterclockwise.

(d) Hold the compass in the centerhold position and rotate your body until the north seeking arrow is under the short luminous line.

(e) Step off by following the line indicated by the two luminous sighting dots on the compass cover.

NOTE: Do not follow the north seeking arrow.

i. Selecting Steering Marks For Night Use.

(1) The general characteristics of a good steering mark for daylight navigation apply to steering marks selected for night navigation. However, as darkness approaches, keep the following factors in mind when selecting steering marks:

(a) Colors disappear at night and objects appear as black or gray silhouettes. If you select a steering mark during daylight because of its distinctive color, it will be of little value as darkness sets in.

(b) During darkness, steering marks must be closer than during daylight.

(c) During darkness, steering marks must have a distinctive silhouette. The silhouette of trees, bushes, and similar objects will change because you see them from slightly different angles as you move up or down hills and when you are detouring obstacles. This applies during daylight also, but to a lesser degree.

(2) Following An Azimuth Without Steering Marks. At times it will be so dark that you can see only a few feet in front of you. When this happens, you have to follow your azimuth without steering marks. The only way that you can stay on your azimuth is by frequently looking at the compass. Practice, experience, and patience will enable you to successfully follow your azimuth under such conditions. But the human tendency to drift will make this technique less accurate than if you use steering marks. You will have to overcome this inaccuracy by thoughtfully selecting your route and checkpoints.

j. Execution Of Navigation At Night. The techniques which produce successful navigation at night are basically the same as those employed during daylight. However, special consideration and emphasis must be given to those aspects of navigation which are affected by the limitations imposed by darkness.

(1) Terrain Association. Just as in day land navigation, at night you should be familiar with the terrain that you will be negotiating. If you should be going down hill when you start off and you are going up hill, then something is wrong. You won't be able to use terrain association to the same extent as during the day, so you may have to rely on dead reckoning. Regardless of the situation, maintain an accurate azimuth and pace count.

(2) Bypassing Large Obstacles At Night. Bypassing an unexpected obstacle at night is fairly simple--just use the 90 degree offset technique. To make a 90 degree turn to the right, hold the compass using the centerhold technique; turn until the center of the luminous letter E is under the luminous line (do NOT move the bezel ring). To make a 90 degree turn to the left, turn until the center of the luminous letter W is under the luminous line. This does not require changing the compass setting (bezel ring), and it ensures accurate 90 degree turns.

(3) NEVER use uncovered flashlights during night navigation because it will reduce your "night vision" and possibly disclose your location to the enemy. If a light must be used, it should be a filtered light and should ALWAYS be used under a poncho or similar cover. If your flashlight does not have a filter, cover the lens with a thin layer of mud so that it reveals only enough light to serve your purpose. Many Marines of all ranks hold the mistaken belief that because they have a red lens on their flashlight, they can not be seen by the enemy. THEY ARE WRONG! They can be seen HUNDREDS of meters away.

(4) Checkpoints. Checkpoints must be easy to recognize. You should be familiar with the characteristics of the checkpoint itself and the terrain features in its vicinity. You should know when you are in the vicinity of a checkpoint. Although not always possible, a checkpoint for night navigation should be of such a nature that missing it entirely is nearly impossible even though you are slightly off course. For this reason a line checkpoint is much superior to a point checkpoint for navigating at night.

9. DETERMINE YOUR LOCATION.

a. Orient A Map. Your first step when navigating in the field is to orient your map. A map is oriented when its north and south correspond to the north and south on the ground. There are two ways of orienting your map - through the use of a compass and by terrain association.

(1) Using A Compass. When orienting a map with a compass, remember that compasses measure magnetic azimuths. Since the north-seeking arrow of the compass points to magnetic north, pay special attention to the declination diagram. Use the following technique to orient your map:

(a) With the map flat on the ground, place the straightedge (on the left side of the compass) along the magnetic north arrow on the declination diagram so that the cover of the compass is pointing toward the top of the map. This will put the fixed black index line of the compass parallel to the magnetic north arrow of the declination diagram (see figure 1201-30).

(b) Keeping the compass aligned as directed above, rotate the map and compass simultaneously until the north-seeking arrow is below the fixed black index line on the compass. Your map is now oriented.

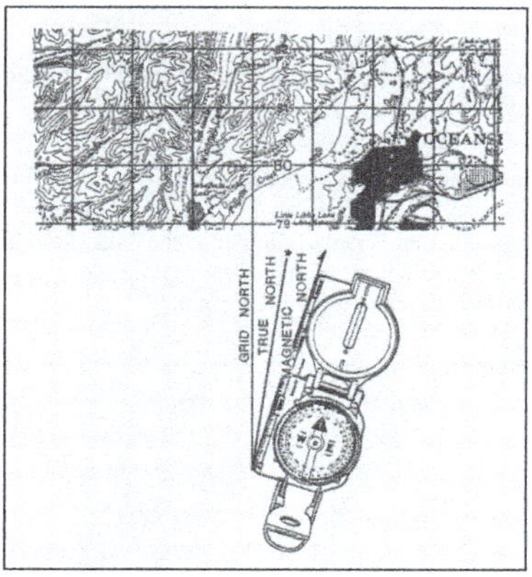

Figure 1201-30

(2) Using Terrain Association. You can orient your map using terrain association when a compass is not available or when you have to make quick references as you move across country. Using this technique requires careful examination of the map and the features on the ground.

(a) Identify Terrain Features. Identify prominent terrain features on the map that you can find on the ground.

(b) Align Terrain Features With The Map. If there is a tower to your right front, then orient the map so that the tower is to your right front. If there is a road off to your left, then ensure the road on the map is parallel to the road on the ground. Once all of the features are lined up, your map is oriented.

b. Determining Your Position. Now that you can orient your map, you must find your location on a map. If you know your approximate location on a map, a study of nearby terrain features will help you determine your position.

(1) Determining Your Location By Inspection. You are standing in the vicinity of several prominent features which can easily be located on the map. By orienting the map and estimating your relation to these features, you should have no difficulty in determining your location.

(2) Determining Your Location By One-Point Resection. One-point resection is an accurate technique of determining your location when you are on or near a linear feature that you can identify both on the ground and on the map. You must also be able to identify another prominent feature, both on ground and on the map. To determine your location by one-point resection follow these steps:

(a) Identify the linear terrain feature that you are located on or near in respect to the ground on your map.

(b) Identify a prominent feature on the ground and locate that feature on your map.

(c) Using the compass-to-cheek technique, sight in on the feature and read the magnetic azimuth.

(d) Convert the magnetic azimuth to a grid azimuth.

(e) Convert this grid azimuth to a grid back azimuth.

(f) With your protractor, plot this grid back azimuth from the feature on the map and extend it until it crosses the linear feature.

(g) Conduct a map inspection to verify your resection.

When selecting a terrain feature, choose one that is perpendicular to the axis of the linear terrain feature so that when you plot the back azimuth on the map, the line will cross the linear feature more or less at a right angle. Figure 1201-31 illustrates the result of a good selection and a poor selection.

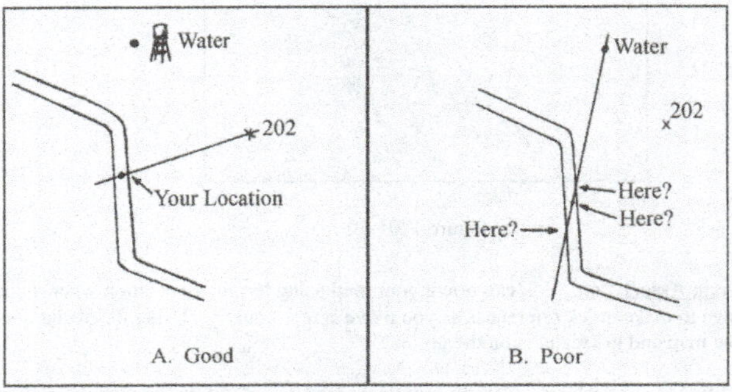

Figure 1201-31

NOTE: A one-point resection can still be done if you are not along a linear feature. Merely estimate the distance between you and your prominent terrain feature. Plot this distance along the back azimuth to get your approximate location.

(3) Determining Your Location By Two-Point Resection. Usually you will find that you are not located on or near a prominent linear feature. Since the accuracy of a one-point resection under these conditions depends on your ability to accurately estimate distance, it is better to use a two-point resection. The procedures for two-point resections are basically the same as for one-point resections except you must select two features instead of one. The back azimuths from each feature is determined and plotted on your map. You are located at the point where these lines cross. If you have a compass and a protractor then follow these steps.

(a) Select two prominent features on the ground whose positions can be located on the map. These features should be at least 30° but not greater that 150° apart (see figure 1201-32).

(b) Using the compass-to-cheek technique, determine the magnetic azimuth to each object.

(c) Convert these magnetic azimuths to grid back azimuths.

(d) With your protractor, draw the respective back azimuths from these two points on your map.

(e) Extend the azimuth lines from these two points until they intersect. You are located at the point where these two lines cross.

(f) Conduct a map inspection to verify your position.

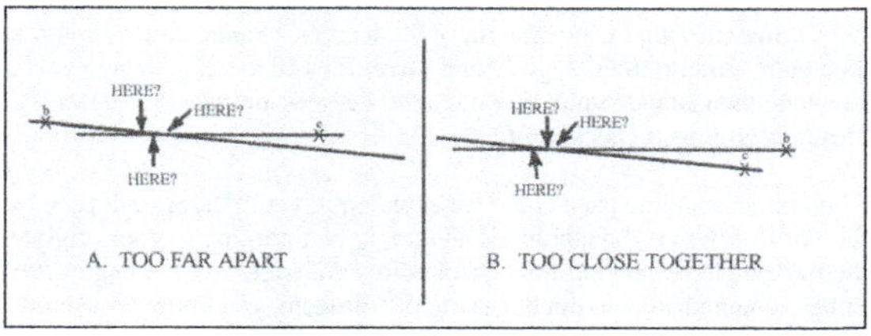

Figure 1201-32

10. BY-PASSING OBSTACLES. Despite the amount of care you take in determining routes, choosing steering marks, and using your compass to maintain a straight line of march, unexpected obstacles will often be encountered which will force you to detour from your intended line of march. There are several ways to bypass obstacles.

a. Detouring Small Obstacles. If you encounter small obstacles, such as trees and bushes, go completely around the obstacle to the center of the opposite side and continue your movement.

b. Detouring Large Obstacles Using Steering Marks. A large obstacle, such as a pond or impassable swamp, can be successfully bypassed if there is a clearly defined steering mark on the far side of the obstacle. Decide carefully whether the object will serve as a suitable steering mark. Remember that an object may have a distinguishing characteristic when observed from a distance, but lose its identity as it is approached. As a final precaution, always leave some mark, such as a small pile of rocks or a blaze (distinctive mark) on a tree on your side of the obstacle, so you can return if you become confused while bypassing the obstacle.

EXAMPLE: You have been following an azimuth of 212 degrees toward your next checkpoint when you come upon an impassable swamp. On the far side of the swamp you see a large pile of rocks that is on a 212 degree azimuth (your original line of march). You can see that this is the only pile of rocks in the area. (If there were others, you would have to carefully study the characteristics of each rock pile and select several features, such as size, color, and nearby objects, which would easily distinguish it from the other rock piles.) Estimate the distance across the swamp, leave a mark on your side of the swamp, put your compass away, and proceed to the rock pile by the easiest route. Upon reaching it you may continue toward your checkpoint along a 212 degree azimuth. (Remember to deduct the estimated distance across the swamp from the remaining distance on that leg.)

c. Detouring Large Obstacles Using Back Steering Marks. If there is no suitable steering marks on the far side of the obstacle, you can bypass the obstacle if there is an easily distinguishable object on your side which can be used as a back steering mark. If there isn't a suitable natural steering mark, then make one. Carefully examine the terrain and shoreline along your line of march on the far side of the obstacle (this will help you estimate when you are at the correct position). Now, estimate the distance across the pond, put away your compass, and proceed to the other side by the easiest route. When you think you are back on your line of march, look back across the obstacle for your back steering mark. Now break out your compass and shoot an azimuth to the back steering mark. If you are in the right place, the mark will lie on the back azimuth of your original azimuth. If it is not, you have not returned to your original line of march. You must move to your right or left until you find the spot from which the mark lies at the correct back azimuth. When you have found this spot, you have returned to your original line of march and may resume your original azimuth. (Again, don't forget to deduct the estimated distance across the pond from your remaining distance on that leg.)

d. Detouring Large Obstacles Using The 90 Degree Off-Set Technique. If you are navigating at night or there is absolutely no way to use steering marks, you can still stay on track using this technique.

(1) Detour the obstacle by moving at right angles (plus or minus 90 degrees) for specified distances. For example (see figure 1201-33), while moving on an azimuth of 90 degrees, change your azimuth to 180 degrees and travel for 100 meters; change your azimuth to 90 degrees and travel for 150 meters; change your azimuth to 360 degrees and travel for 100 meters; then change your azimuth to 90 degrees and you are back on your original azimuth line, 150 meters closer to your next checkpoint.

(2) You must maintain an accurate pace count for each leg of your detour. All paces which are in the direction of your original line of march must be combined so that, upon returning to your original line of march, you know the distances yet to be traveled to reach your next checkpoint. All paces at right angles (90 degrees) to your original line of march must be computed so that, upon passing the obstacle, you know how many paces it will take to return you to your original line of march. You should return to the original line of march as soon as possible after passing the obstacle. The only sure way of knowing that you are back on this line is by maintaining an accurate pace count.

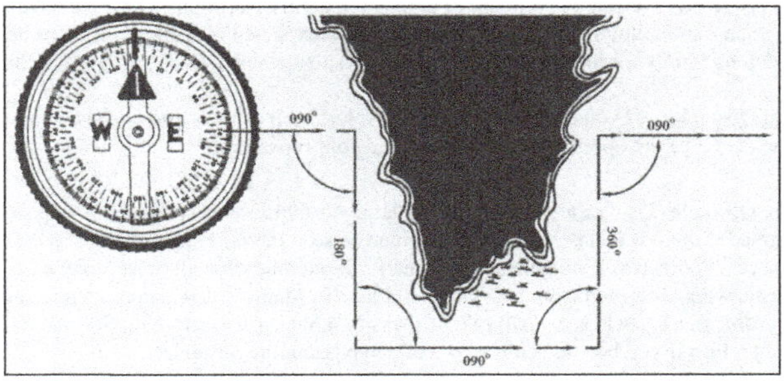

Figure 1201-33

APPENDIXES:

A. Example of Land Navigation Evaluation Checklist
B. Sample of Practical Application

REFERENCE:

1. FMFM 21-26, Land Navigation

APPENDIX A

SAMPLE PE CHECKLIST

LAND NAVIGATION PERFORMANCE EVALUATION CHECKLIST

Learning Objectives.

1. Without the aid of references and given a map, protractor, and eight (8) digit grid coordinates, plot the points on the map in order to reinforce your knowledge and skills in navigating on land, per FM21-26. (SCRS12.01.01b)

2. Without the aid of references and given a map, protractor, and eight (8) digit grid coordinates, measure the distance from one point to another in order to reinforce your knowledge and skills in navigating on land, per FM21-26. (SCRS12.01.01c)

3. Without the aid of references and given a map, protractor, and eight (8) digit grid coordinates, determine the grid azimuth between points in order to reinforce your knowledge and skills in navigating on land, per FM21-26. (SCRS12.01.01d)

4. With the aid of references and given a checklist, area of operations, lensatic compass, map of the area, eight (8) digit grid coordinates, starting point, finishing point, and as a member of a two (2) man team, navigate during daylight hours in order to locate five (5) checkpoints on a map of the area in a time limit of four (4) hours in order to reinforce your knowledge and skills in navigating on land, per FM21-26. (SCRS12.01.01h)

5. With the aid of references and given a checklist, area of operations, lensatic compass, map of the area, eight (8) digit grid coordinates, starting point, finishing point, and as a member of a two (2) man team, navigate during darkness hours on order to locate three (3) checkpoints on a map of the area in a time limit of four (4) hours in order to reinforce your knowledge and skills in navigating on land, per FM21-26. (SCRS12.01.01I)

Instructions To The Student.

CLASSROOM PORTION. "For this class, you will be required to plot a total of (10) eight digit grid coordinates, measure the distance between each point, determine the grid azimuth between the points, and convert the grid azimuth to a magnetic azimuth. You will be graded by a checklist in accordance with FM 21-26. You will have 30 minutes to complete this task. If you have questions feel free to ask me, your time begins now.... Good luck."

FIELD PORTION. "For the day land navigation, I will show you what the start-points and checkpoints look like. Next we will move over to the 100 meter pace course, so you can determine your pace count for 100 meters. At this time listen up for your safety brief, Instructor Note: (in accordance with your Academy SOP). You will have 4 hours to locate five points. The time on deck is now _____ , your drop dead time is ______. Your drop dead time means, the time that you will be back here inside the CP. Your escape azimuth is ____________. Good Luck!!"

"For the night land navigation, we will move over to the 100 meter pace course, so you can determine your pace count for 100 meters at night. At this time listen up for your safety brief, Instructor Note: (in accordance with your Academy SOP). You will have 4 hours to locate three points. The time on deck is now _____ , your drop dead time is ______. Your drop dead time means, the time that you will be back here inside the CP. Your escape azimuth is ____________. Good Luck!!"

Land Navigation
Performance Evaluation Checklist

CARD # _________

NAME: ____ PLATOON: ____________________

NAME: ____ GM ANGLE IS ___________ DEGREES

DAY STARTING POINT # ______

STARTING POINT: __________________ (plotted Correctly **1 point**) ______

		1 Point Each	**1 Point Each**		**1 Point Each**	
Plot Point	Grid	Plotted Correctly	Grid Azimuth	Magnetic Azimuth	Distance	Total Points
#1						
#2						
#3						
#4						
#5						
					Total Points	

NIGHT STARTING POINT # ______

STARTING POINT: __________________ (plotted Correctly **1 point**) ______

		1 Point Each	**1 Point Each**		**1 Point Each**	
Plot Point	Grid	Plotted Correctly	Grid Azimuth	Magnetic Azimuth	Distance	Total Points
#1						
#2						
#3						
					Total Points	

Day Point Identification Answers	**8 Points Each**
Point # 1:	
Point # 2:	
Point # 3:	
Point # 4:	
Point # 5:	
Total Points	

Night Point Identification Answers	**8 Points Each**
Point # 1:	
Point # 2:	
Point # 3:	
Total Points	

Day Time: ______ Time Points (0 or 5) ________ Night Time: ______ Time Points (0 or 5) ________

Total Points Land Nav: _______ (Final Grade)

Student's Signature: ____________________________________

Evaluator's Signature: __________________________________

APPENDIX B

SERGEANTS COURSE
SAMPLE PRACTICAL APPLICATION

1. Locate and identify the most prominent terrain or man made feature for the following grid location.

    a. 93326721_____________________

    b. 89927653_____________________

    c. 91556780_____________________

    d. 92567314_____________________

    e. 89307351_____________________

2. Determine the straight-line distance between the following points.

    a. Garfield High School (9979) and Bel Air (9879).
    ___________ meters

    b. The intersection of route 640 and 234 (9078) and Bench Mark 101.9 (9177).
    ___________ meters

    c. Greenwood Church (9578) and Neabsco Church (9876).
    ___________ meters

3. Determine the irregular map distance between the following points.

    a. The intersection of route 640 and route 234 (9078) and the intersection of route 646 and route 234 (8878) along route 234.
    ___________ meters

    b. Bench Mark 124.0 (8779) and Lookout Tower (8781).
    ___________ meters

    c. The intersection of interstate 95 and route 610 (9975) and intersection of interstate 95 and route 234 (9772) along interstate 95.
    ___________ meters

4. Plot the following azimuths with your protractor and answer the questions in the space provided below.

    a. From Garfield High School (9979), plot a grid azimuth of 91 degrees. What topographical feature is located 2000 meters away along this azimuth?

    ____

    b. From LZ Duck (8274), plot a gird azimuth of 232 degrees. What topographical feature is located 5440 meters away along this azimuth?

    ____

    c. From Cole School (8881), plot a grid azimuth of 195 degrees. What topographical feature is located 2050 meters away along this azimuth?

____

5. Resection (8 digit grid is required)

   a. You are moving NNE along an unimproved road located in grid 7078. You take a azimuth of the lake toward the East of 105 degrees (magnetic), and another of the lake toward the South of 211 degrees (magnetic). What is your position? ________________

   b. You are located somewhere 9167, you take an azimuth of 30 degrees (magnetic) to a dam that is 610 meters away. What is your located? ________________

   c. You are moving South along an unimproved road located in grid 9667. You take an azimuth SSW to Ashrust School, which yields 229 degrees (magnetic), and another to the SE to a water tower which yields 146 degrees (magnetic). What is your position? ________________

Made in the USA
Monee, IL
05 November 2026

67808354R00085